Studies in the Foundations
Methodology and Philosophy of Science 4

Problems in the Foundations of Physics

Edited by
Mario Bunge

With 3 Figures

Springer-Verlag
Berlin Heidelberg New York 1971

Mario Bunge McGill University, Montreal/Canada

ISBN 3-540-05490-1 Springer-Verlag Berlin · Heidelberg · New York
ISBN 0-387-05490-1 Springer-Verlag New York · Heidelberg · Berlin

Typesetting, printing and binding: Universitätsdruckerei H. Stürtz AG, Würzburg

Preface

This is a collection of technical papers in the foundations and the philosophy of physics with emphasis on the former. It takes both "foundations" and "philosophy" in their narrow technical senses but it construes "physics" *lato sensu*, as including all the sciences of nonliving systems. All eleven papers constituting this volume were written for it.

The problems tackled in this book concern certain basic concepts, hypotheses, theories, and research programmes in physical science. Some of these problems are topical, others new, but they are all fundamental and the subject of research and controversy.

Consequently this volume is expected to serve those students, teachers and researchers who enjoy learning, teaching, discussing or doing theoretical physics. It is addressed to the nine to niners rather than to the nine to fivers. It is expected to attract the theoretician in search for new basic ideas, the teacher eager to perfect his understanding of physical theory and transmit his own zeal and his own doubts, as well as the student anxious to get down to essentials. This book may also interest the mathematician for whom physics offers a challenge (or a good pretext). Finally, it should get the attention of the philosopher of science aware of the advantages of philosophizing on foundations research problems rather than on the popularization of some results of research.

There are at least two reasons for valuing foundations research. One is purely intellectual and aesthetic: We like to know not only what is being done but also how it is being done and why, and how it might be improved on. The more we love a subject the greater pleasure we derive from working at it and beholding it. And conversely: The better we know the subject the more intensely we are bound to love it. Although most of us are carpenters and plumbers rather than architects or master builders, if we are skilled and happy craftsmen we shall take pleasure and pride in contemplating the building as a whole, from its changeable foundations upwards.

A second reason for valuing the foundations and philosophy of science is of a practical nature. In fact these disciplines take part, usually in a covert way, in the planning of research, in the allocation of research facilities and funds, and in the design of teaching curricula. And any such decision may affect entire generations of scientists. For example, the indiscriminate production of highly specialized scientists bolted to a given computation method or a given measuring device is likely to contribute to unemployment – and to boredom. A balanced diet, with emphasis on general and basic theories as well as portable methods, is likely to produce more useful investigators – and happier human beings, for whom science is both job and hobby.

May this volume, for focusing on fundamentals, wide range issues, versatile tools, and long term research programmes, contribute to reinforcing the long view of physical theory.

MARIO BUNGE

Foundations and Philosophy of Science Unit
McGill University,

Montreal, Spring 1971

Contents

I. General

1 A Mathematical Theory of the Dimensions and Units of Physical Quantities

MARIO BUNGE

Foundations and Philosophy of Science Unit
McGill University, Montreal (Canada)

The aim of this investigation is to disclose the mathematical status of the dimensions and units of physical quantities. Every physicist knows how to perform a dimensional analysis and how to calculate the units of a dimensionally complex magnitude, but no theory proper has hitherto justified these operations. Worse: the two concepts are often mixed up. Even worse: they are frequently "defined" by reference to measurement operations. In this paper an algebra of dimensions and an algebra of units are built with elementary set theoretic and algebraic resources. By supplying unambiguous and exact elucidations of the concepts of dimension and of unit, those theories may contribute to clearing up some of the confusions that becloud the subject.

1 Dimensions

1.1 Introduction. Dimensional analysis is an established branch of the foundations of physics. However, it is more of a loose collection of concepts, theorems and rules than a hypothetico-deductive system. It is much in the same state as HEAVISIDE'S "operational calculus" before it was legitimated by the theory of Laplace transforms. So much so that dimensional analysis does not even supply a mathematical analysis of the very notion of a dimension. Consequently we operate "symbolically", i.e. *de facto* but not *de jure*, with dimensions without quite knowing what they are. We thereby espouse, tacitly to be sure, the pragmatist motto "Things are what they do". This attitude, suitable as it may be for gadgets, is dangerous with regard to ideas.

In this section we provide a justification for the calculus of dimensions. The basic idea, first adumbrated in [1], is this. Consider the set Q of all physical quantities: durations, entropies, state functions, dimensional constants, and so on and so forth. These concepts sort themselves out into dimension species such as the class L of all length-like quantities: heights, depths, widths, wavelengths, mean free paths, etc. In other words, there is a certain equivalence relation $\equiv$ that induces a partition of the totality Q of physical quantities into homogeneous subsets (species) so that, if q and q' are in Q, then: $q \equiv q'$ if and only if q and q' are of the same species, say L(for length-like), or T(for time-like), or 1_D(for dimensionless). The relation $\equiv$ of equidimensionality is determined by a certain function on Q that assigns every physical quantity its proper dimensional niche. This function, usually designated by [ ], will be called the dimension function. The value of this function at a physical quantity is precisely some dimension species. For example, the dimension value of a wavelength λ is the dimension species L, i.e. $[\lambda] = L$. What follows is a systematic elaboration of this idea.

1.2 Physical Quantity. Every physical theory is a set of formulas built with certain basic functions and constants called physical quantities. At least one of these must be a magnitude, i.e. a function concerning, or referring to, one or more physical systems and taking values in a subset of real numbers or in a family of intervals of real numbers. (A complex valued function referring to some physical entity will count as two magnitudes and every component of a tensor may be regarded as one magnitude provided it concerns some physical system or other.) In other words, any function $P\colon A \times B \times \cdots \times N \to V$, where V is included in the real line R or in its power set $\mathbb{P}(R)$, is a *magnitude* provided at least one of the cartesian factors A through N is a kind of physical system, e.g. the set of all magnetic fields, or the set of all Carbon molecules. This characterization of a magnitude will suffice here. A more precise definition will be introduced in Section 2.2 in relation with the concept of a unit.

Call Q_T^* the set of basic (undefined) physical quantities in a theory T, with the understanding that Q_T^* includes a nonempty set of magnitudes, or quantities with a factual (e.g. physical) referent. With the help of mathematics any number of derived (defined) physical quantities can be built out of Q_T^*. Call Q_T the infinite set of physical quantities, basic or derived, in a theory T. (Q_T has at least the group structure, but this will not concern us here.) And call Q an arbitrary set of physical quantities: either a subset of a certain Q_T or the union of any number of Q_T's.

In addition to its peculiar set Q_T^* of basic physical quantities, every physical theory T contains a set D_T^* of basic dimension values such as L, T, and M. (There is no direct relation between Q_T^* and D_T^*: the asterisk indicates only that the elements of these basic sets are not definable within T.) Loosely speaking D_T^* is the generator of the totality D_T of dimension values of T. This enlarged set D_T includes all the rational powers of the elements of D_T^* as well as their finite concatenations, e.g.

$L^{\frac{3}{4}} T^{\frac{8}{5}} M^{-\frac{2}{9}}$. Our first task will be to tame D_T. Later on we shall investigate the manner in which each Q_T is associated to its D_T.

1.3 Dimension Value. We start by laying down

Axiom 1. For every set Q of physical quantities there is a finite set $D_Q^* = \{A, B, \ldots, N\}$ of independent elements called the *basic dimensions values* associated with Q.

Next we postulate that the basic dimension values can be raised to rational powers as if they were real numbers. That is, if $X \in D_Q^*$ and if m and n are natural numbers, then

$$
\begin{array}{ll}
X^0 = 1_D & X^{-1} \cdot X = X \cdot X^{-1} = 1_D \\
X^{n+1} = X^n \cdot X & X^{-n} = (X^n)^{-1} \\
X^{m+n} = X^m \cdot X^n & X^{-m-n} = X^{-m} \cdot X^{-n} \\
(X^m)^n = X^{mn} & (X^m)^{-n} = X^{-mn} \\
Y = X^{1/n} =_{\mathrm{df}} Y^n = X & X^{-1/n} = (X^{-1})^{1/n} \\
X^{m/n} = (X^m)^{1/n} & X^{-m/n} = (X^{-m})^{1/n}.
\end{array}
$$

We compress the preceding formulas into

Axiom 2. Let X be a basic dimension value. Then the structure

$$\mathscr{D}_X = \langle D_X, \cdot, ^{-1}, 1_D \rangle,$$

where $D_X = \{X^\alpha \mid \alpha \text{ rational}\}$, $\cdot$ a binary operation on D_X, $^{-1}$ a unary operation on D_X, and 1_D a distinguished element of D_X, is an *abelian group*, written multiplicatively, in which the powers are defined just as in the field of reals.

The basic set of this group, i.e. D_X, is constituted by all of the rational powers of the dimension value X. It deserves a special name as assigned by

Definition 1. Let $X \in D_Q^*$. Then the set $D_X = \{X^\alpha \mid \text{rational}\}$ is called the *dimension genus X*.

1.4 System of Dimension Values. We now extend D_X to the set of all dimension values, basic and derived, formed from a given set D_Q^* of basic dimension values. That is, we proceed to sanction operations such as $X^\alpha \cdot Y^\beta$, where X and Y are dimension values, and α, β rationals. We do this by means of

Axiom 3. Let Q be a set of physical quantities and let D_Q^* be the associated set of basic dimension values. Then the structure $\mathscr{D}_Q = \langle D_Q, \cdot, ^{-1}, 1_D \rangle$, where $D_Q = \{A^\alpha \cdot B^\beta \cdot \cdots N^\nu \mid A, B, \ldots, N \in D_Q^*$ and $\alpha, \beta, \ldots, \nu$

rationals}, $\cdot$ a binary operation on D_Q, $^{-1}$ a unary operation on D_Q, and 1_D a selected element of D_Q, is an *abelian group* written multiplicatively.

In other words, the infinite set D_Q of all dimension values associated with any given set Q of physical quantities is assumed to have the group structure. We call it the *system of dimension values of Q*.

Let us now find out the relations between the dimension genera D_X (Definition 1) and the totality D_Q of dimension values. Clearly, every one of the former is included in the latter. More precisely, we have

Theorem 1. Let Q be a set of physical quantities and let D_Q be its set of dimension values. Then, for every $X \in D_Q$, $D_X \subseteq D_Q$.

Furthermore every D_Q can be split into as many factors as basic dimension values. More exactly, we shall prove

Theorem 2. Let Q be a set of physical quantities, D_Q its associated set of dimension values, and $\mathscr{D}_X = \langle D_X, \cdot, {}^{-1}, 1_D \rangle$ the group of dimension values of genus $X \in D_Q$. Then the system $\mathscr{D}_Q$ of dimension values of Q is the direct product of the dimension genera: $\mathscr{D}_Q = \mathscr{D}_A \times \mathscr{D}_B \times \cdots \times \mathscr{D}_N$.

Proof. By Axiom 2 every factor is an abelian group and by Theorem 1 it is included in D_Q. By Axiom 3 the products are commutative. Finally the only element all n factors share is 1_D. So the conditions for a direct product of groups are met.

So far, a dimension value is just an element of a certain group associated with a set of physical quantities. The nature of the elements of this group will become clearer as we find out the kind of association they hold with Q, a task we now turn to.

1.5 Dimension Function. We introduce the dimension function, relating Q to D_Q, through

Axiom 4. Let Q be a set of physical quantities and D_Q the system of dimension values of Q. The *dimension function* $[\ ]: Q \to D_Q$ assigns each physical quantity $q \in Q$ a dimension value $d \in D_Q$ in such a way that

(i) $[1] = 1_D$,

(ii) if p, $q \in Q$, and α, β are rationals, then

$$[p^\alpha \cdot q^\beta] = [p^\alpha] \cdot [q^\beta].$$

This postulate entails all the desired formal properties of the dimension function. Among them those stated in the following

Corollary. If p and q are physical quantities, then

(i) $[p \cdot q] = [p] \cdot [q]$

(ii) $[p \cdot q^{-1}] = [p] \cdot [q]^{-1}$.

The presystematic notion of dimension has now been split into the clear cut concepts of dimension function and dimension value. And these will in turn help us define a few other standard concepts such as those of dimensionless quantity and dimensional constant:

Definition 2. Any $q \in Q$ for which $[q] = 1_D$ is called a *dimensionless quantity.*

Definition 3. A scalar with a dimension value other than 1_D is called a *dimensional constant* or *scale factor.*

A possible partition of a set of physical quantities is into dimensional and dimensionless. A second partition is into dimensional constants (a subset of dimensional quantities) and magnitudes. While dimensional constants concern no physical system (they have no physical referents), magnitudes are associated with definite kinds of physical system. (Recall Section 1.2.) Thus the gravitational constant is a dimensional constant not a property of a physical system, and this both in the classical theory of gravitation and in EINSTEIN'S general relativity. A third possible partition of a set of physical quantities is into dimension genera. Indeed, the Definition 1 of dimension genus D_X and Axiom 4 allow us to introduce an equivalence relation $\sim$, interpretable as equivalence in respect of dimensional genus, namely thus:

Definition 4. If $p, q \in Q$, then:

$$p \sim q \quad \text{iff} \quad [p], [q] \in D_X .$$

For example, $L^{\frac{1}{2}} \sim L^{-\frac{3}{7}}$. But we are more interested in a different, far more natural partition of Q, which shall be introduced presently.

1.6 Equidimensionality and Physical Quantity Species. Consider the set Q of all physical quantities, basic and derived, occurring in all possible physical theories. In order to class them according to their dimension value, i.e. to partition Q into dimension species, we need the equidimensionality relation. This relation, which shall be designated by $\equiv$, is generated by the dimension function [ ], introduced by Axiom 4, according to the following

Definition 5. If $p, q \in Q$, then p and q are said to be *equidimensional* iff they have the same dimension value, i.e.

$$p \equiv q \quad \text{iff} \quad [p] = [q] .$$

We can easily prove the following trivial

Theorem 3. The equidimensionality relation $\equiv$ is an equivalence relation [i.e. it is reflexive, symmetrical and transitive] on Q.

In other words, $\equiv$ is the equivalence kernel of [ ] and as such it is the relation competent to group physical quantities into dimensionally homogeneous species. Every such class is an equivalence class: we shall call it a dimension species. More explicitly, we stipulate

Definition 6. If $q \in Q$, then the equivalence class of all the $\equiv$ relatives [or species partners] of q is the *dimension species* of q.

By a happy coincidence the standard notation for the equivalence class generated by an element $q \in Q$ is $[q]$, the same as for the dimension value of q. We shall prove the identity of these two concepts. But to this end we must take a deeper look at the dimensional partition of Q.

1.7 Dimension Value and Dimension Species. Consider the natural dimensional partition of Q, i.e. the partition of Q induced by the equidimensionality relation $\equiv$. This is the quotient set of Q by $\equiv$, i.e. $Q/\equiv$, the *family of all physical quantity species.* There is a function that assigns each q in Q the equivalence class $[q]$ in $Q/\equiv$. This function is none other than the dimension function. In other words, we have

Theorem 4. The dimension function [ ] is the projection of Q onto the dimensional partition of Q, i.e.

$$[\ ]: Q \rightarrow Q/\equiv, \quad \text{with} \quad [q] = \{p \mid p \in Q \text{ and } p \equiv q\}$$

or, equivalently, $[q] = p_{\equiv}\, q$, where $p_{\equiv}$ denotes the projection map.

The proof is easily gotten with the help of [2], p. 21. The last theorem, jointly with Axiom 4 and the definition of a function, entail the following

Theorem 5. The set of dimension values is equal to the family of all dimension species:

$$D = Q/\equiv .$$

Finally, by the principle of extensionality of set theory, we obtain

Theorem 6. If $q \in Q$, then the dimension value of q is the same as the dimension species of q.

The last theorem justifies using the same symbol, namely $[q]$, to designate both the dimension value and the dimension species of a physical quantity q.

This winds up our axiomatic theory of the dimensions of physical quantities. Let us see whether it is of any use in assigning dimension values—a practical goal of dimensional analysis.

1.8 The Assignment of Dimension Values. Our theory does not replace the usual dimensional analysis but intends to give it a foundation. The function of our theory is indeed only to elucidate the general notions of dimension in precise mathematical terms. It performs this task without committing itself as to the memberships of the sets Q and D. Consequently our theory is insufficient to determine the dimension value of every physical quantity. Axiom 4 and its logical consequences will determine the dimension of a compound physical quantity, such as the Coulomb force, provided the dimensions of its ingredients (electric charge, dielectric constant, and mutual distance) be given.

Now, assigning specific dimension values to simple or unanalyzed physical quantities is a task for specific postulates. In most cases this is felt to be unnecessary because we can recognize intuitively the dimension of a basic familiar quantity. But in some cases (e.g. the dielectric constant) there are ambiguities: different dimension values can be postulated and correspondingly slightly different theories will be possible. In such cases an explicit dimension assignment will be mandatory. Once such an assignment has been made for the unanalyzed quantities, Axiom 4 will yield the dimension of any multiplicatively composed quantity.

We may thus jump to the following general prescriptions.

Rule 1. In an axiomatic reconstruction of a physical theory, all those and only those basic (undefined) physical quantities that are not analyzed in that theory should be assigned a dimension value by postulate.

Rule 2. In an informal presentation of a physical theory, only the dimensional constants (e.g. the Boltzmann constant) and those magnitudes for which dimensional ambiguities may arise need be assigned dimension values by explicit hypothesis.

1.9 Concluding Remarks. We have proposed a mathematical theory of the dimensions of physical quantities based on the concepts of physical quantity, dimension value, and dimension function. The formal properties of the last two concepts have been postulated so as to retrieve the usual rules of thumb of the calculus of dimensions. Every dimension value is a set and more precisely an equivalence class. And every system of dimension values is a commutative group. The dimension function [ ] defines an equivalence relation $\equiv$ interpretable as the relation of equidimensionality. This relation allows us to improve on the initial characterization of a dimension value: it is also a species belonging to the natural dimensional partition $Q/\equiv$.

The veil of mystery that shrouds the intuitive concept of a dimension has thus been lifted, revealing a modest but neat mathematical structure. The question "What are dimensions?" is seen to be ill conceived, for there are two concepts of dimension, namely those of dimension function and dimension value. The correct question is: "What are [ ] and d?" The answer is just this: "They are the mathematical objects satisfying the theory expounded above." Admittedly this answer fails to give us a "feel" for dimensions. But the same holds for every other basic concept: While theory alone supplies exactness, only practice breeds familiarity.

To conclude this section, note that we have made no mention of other subjects that are often mixed up with dimensions—notably units, scaling, and measurement. The relation of dimension values to units will be studied in the next Section.

2 Units

2.1 Introduction. The concept of a unit is one of those scientific notions that has remained obscure for want of a theory and excess of a coarse philosophy. Indeed we know how to manipulate units but we do not seem to know what kind of mathematical object they are. As a consequence units are sometimes confused with dimensions, at other times with standards, and they are often introduced in relation to measurement rather than to concept formation. Being a theoretical concept, the notion of a unit should be elucidated within a theory proper. And being a generic notion, i.e. one occurring in every branch of quantitative science, its elucidation through a mathematical theory is a task for the foundations of science. What holds for units holds for any expression containing them, such as '3 sec' and '$1\ \text{barn} = 10^{-24}\ \text{cm}^2$'. Such "symbolic" expressions make no sense at all unless they are embedded in definite algebraic structures sanctioning the usual operations among units and among units and numbers. Let us investigate these structures.

2.2 Units and Coherent Systems of Units. Our starting point must be the concept of a magnitude, for the values of a magnitude are reckoned on some system of units and, conversely, a unit exists only by virtue of its association with a set of magnitudes. (Example: No length value without a length unit and no length unit without a length concept.) We can accomplish this goal by adopting the analysis given in [1], which can be summarized in the following.

Axiom 5. Let $A, B, \ldots, N$ be n kinds of physical system, where n is a positive integer. And let R designate the real line and $\mathbb{P}(R)$ the power set of R. Then there exists a nonempty set U_P, called the set of *P-units*, and there is at least one function

$$P: A \times B \times \cdots \times N \times U_P \to V, \quad \text{where } V \subseteq R \text{ or } V \subseteq \mathbb{P}(R),$$

called a *magnitude*, such that P represents a (mutual) property of every n-tuple $\langle a, b, \ldots, n \rangle$ with $a \in A, b \in B, \ldots, n \in N$.

In plain words: every physical system, whether simple or complex, is characterized by properties representable as real functions, and the values of these functions depend not only on the system itself but also on the units that have been adopted conventionally. For example, in nonrelativistic mechanics mass is represented by a certain function $M: B \times U_M \to R^+$, where B is the set of bodies and U_M the set of mass units. The precise properties of that function M are given by the postulates in which M occurs.

Let us now get a step closer to the conventional ingredient in a magnitude. We first stipulate

Definition 7. Let M be a set of magnitudes. Then the set U_M of *units for* M is the union of the sets of X-units for $X \in M$, i.e.

$$U_M = \bigcup_{X \in M} U_X .$$

Now, units can be basic (simple) or derived (compound). Thus cm is simple and $\mathrm{cm}^2\,\mathrm{sec}^{-1}$ is compound. And given any set of units there are many ways of analyzing them into simpler units. In short, there are many systems of basic units, such as {cm, g, sec} and {ft, lb, day}. We consecrate this fact of life, or rather convention, by postulating

Axiom 6. Let M be a set of magnitudes and U_M the set of units for M. Then there is at least one nonempty finite subset

$$U_M^* = \{u_1^*, u_2^*, \ldots, u_m^*, 1_U\} \subset U_M ,$$

called a *system of basic units for* M, such that

(i) the elements of U_M are mutually independent (not interdefinable);

(ii) the structure $\mathscr{U}_M^* = \langle U_M^*, \cdot, {}^{-1}, 1_U \rangle$ is an abelian group written multiplicatively, in which the powers are defined as in the field of reals;

(iii) for every $u \in U_M$ there is a μ-tuple $\langle \alpha, \beta, \ldots, \mu \rangle$ of rational numbers such that u is uniquely decomposable in the form

$$u = u_1^{*\alpha} \cdot u_2^{*\beta} \cdot \cdots \cdot u_m^{*\mu} .$$

Finally we individualize, within the population U_M, the units of a feather, i.e. the coherent systems of units. This task is performed by

Definition 8. Let M be a set of magnitudes and U_M the set of all of its units. Then the structure $\mathscr{U}_M^C = \langle U_M^C, \cdot, {}^{-1}, 1_U \rangle$, with $U_M^C \subseteq U_M$, is called a *coherent system of units* for M if and only if it is an abelian group (written multiplicatively).

In other words, the members of a coherent system of units have inverses and they concatenate freely even if they are not basic units. The belonging to the same coherent system of units, as is the case with $\mathrm{cm}^{\frac{1}{2}}$ and $\mathrm{sec}^{-\frac{3}{7}}$, defines an equivalence relation E on U_M. In turn this equivalence relation induces the partition U_M/E of any set of units into coherent systems. But we shall not pursue this line here.

If we wish to compound units we must choose them from a coherent system. But the injunction to employ only coherent systems of units in any given theory is the subject of a methodological rule that falls outside the algebra of units. Whether or not the rule is implemented, units are uniquely characterized by the above axioms and definitions, which moreover legitimate the "symbolic" operations with units within a coherent system. Our next task will be to legitimate the equally "symbolic" multiplications of numbers by units.

2.3 Denominate Numbers. We shall presently justify the common practice of writing expressions of the form "$r\,u$", where r is some real number (the *numeric* of the magnitude concerned) and u an element of some set U_M of units for the given magnitude. The usual justification is success. Ours will consist in introducing an algebraic structure making room for the totality of expressions of the form "$r\,u$". This structure, to be called an R-group, may be thought of as a merger of a coherent system of units with the field of reals. It must be something more complex than the direct product of the real line (regarded as a group) and a group of units, these two structures do not share the same unit element: $1 \neq 1_U$.

We start by defining an R-group. It constitutes the multiplicative partner of an R-module, itself an additive abelian group equipped with scalar multiplication or, rather, multiplication by the elements of a ring of scalars. More precisely, we stipulate

Definition 9. The structure $\mathscr{R} = \langle R, S, f, \cdot, \circ, ^{-1}, 1_S \rangle$ is an *R-group* if and only if

(i) $\langle R, +, \cdot, 1 \rangle$ is a commutative ring with unit element 1;

(ii) $\langle S, \circ, ^{-1}, 1_S \rangle$ is a multiplicative abelian group with unit element 1_S;

(iii) f is a function from $R \times S$ into S, written $\langle r, s \rangle \mapsto r\,s = s' \in S$, subject to the following conditions:

(a) for every $r \in R$ and any $s, s' \in S$: $r(s\,s') = (r\,s)\,s' = s(r\,s')$;

(b) for any $r, r' \in R$ and every $s \in S$: $(r\,r')\,s = r(r'\,s)$;

(c) for every $r \in R$: $r\,1_S = r$.

Note that, since S is a multiplicative group, the addition of arbitrary elements of it is not defined. Nor do we wish to sanction expressions like "$r\,s + r'\,s'$", which could be interpreted as sums of heterogeneous denominate numbers. Sums of denominate numbers make sense only if a unit can be factored out of them, as with "2 cm + 3 cm". To account for this case we define a new operation $+'$ in the set S by means of

Definition 10. For any $r, r' \in R$ and any $s \in S$,

$$r\,s +' r'\,s = (r + r')\,s.$$

Finally we restrict the function f occurring in Definition 9 to $R \times T$, where T is a subset of S, and call g the new function. Explicitly: we introduce

Definition 11. Let $T \subset S$, where S is the second set occurring in an R-group, and let f be the function involved in the latter. Then $g = f | R \times T$.

We are now ready to apply these concepts to our problem. We make the following assignments:

$R :=$ The real line.

$S :=$ The totality of denominate numbers of the form "$r\,u$" with $r \in R$ and $u \in T = U_M^C$.

$T := U_M^C =$ A coherent system of units for the magnitudes M.

$1_S := 1_U =$ The unit element of U_M^C.

The result is the definition of a particular model of an R-group, that justifies the standard rules governing the "symbolic" (nonarithmetic and theoretically nonjustified) products of numbers by units. We stipulate then

Definition 12. The structure $\mathcal{N}_M^C = \langle R, S, g, \cdot, \circ, ^{-1}, 1_U \rangle$, with $g = f | R \times U_M^C$, is a *coherent system of denominate numbers* if and only if it is an R-group and if R is the field of real numbers and if U_M^C is a coherent system of units for some set M of magnitudes.

The foregoing nearly justifies writing '$P(a) = p\,u$' for the value of the magnitude P at the physical system a on the unit u. Not quite, because we have to do with two different functions: P and what we shall call its associated magnitude P'. More precisely, we stipulate

Definition 13. If $P: A \times B \times \cdots \times N \times U_P \to V$ is a magnitude concerning a complex of physical systems of the kinds A through N, then the *associated magnitude* is the function $P': A \times B \times \cdots \times N \to W$, where $W \subset S$ is a subset of the set of denominate numbers, such that

$$P'(a, b, \cdots, n) = p\,u \quad \text{iff} \quad P(a, b, \ldots, n, u) = p.$$

Definition 14. The numerical value of $P: A \times B \times \cdots \times N \times U_P \to V$ at $\langle a, b, \ldots, n, u \rangle$ with $a \in A, b \in B, \ldots, n \in N$ is called the *numeric* of P' for $\langle a, b, \ldots, n \rangle$.

The extension of the preceding definitions to vector valued and tensor valued functions is quite unnecessary, as they take care of every component.

Our axioms and definitions in this section sanction the otherwise lawless expressions of the form "$r\,u$". They also dispel the popular belief that "3 cm" is of the same kind as "3 planets".

2.4 Species and Changes of Unit. The inch and the micron, though different, belong to the same set U_L of length units, itself a subset of the totality U of units. We may thus write: in $\sim\mu$, where the relation $\sim$ is an equivalence relation on U. More precisely, we lay down

Definition 15. Given a magnitude P and its associated set U_P of units, any two members of this latter set are said to be *congruent*. In symbols,

$$\text{If } u, u' \in U, \text{ then: } u \sim u' \text{ iff } u, u' \in U_p \subset U.$$

The name "congruence" for $\sim$ has been chosen advisedly: $\sim$ is in fact a congruence with respect to the multiplication of units. That is, if $u, u', v, v' \in U_P$, then:

$$\text{If } u \sim u' \quad \text{and} \quad v \sim v', \qquad \text{then: } u \cdot v \sim u' \cdot v'.$$

Since the relation of units congruence is an equivalence relation, it induces a partition of any set U into mutually disjoint equivalence classes, one per magnitude. Every such set is a species of units and the totality of such species, i.e. the quotient set $U/\sim$, may be called the *natural partition* of U. To be more explicit, we lay down

Definition 16. If $u \in U$, the *species* of u is the $\sim$-equivalence class of u. In symbols,

$$[u] = \{v \mid v \in U \text{ and } v \sim u\}.$$

Caution: In the present context [ ] does not stand for the dimension function introduced in Section 1.5.

Definition 17. The quotient set $U/\sim$ is called the *family of species of units*.

Although U is a collection of individuals (units) and $U/\sim$ a family of sets (species of units) they have the same structure. Indeed, one can define an operation $\square$ on the family of species of units as follows:

$$[u] \square [v] =_{\text{df}} [u \cdot v].$$

This operation is associative. It is also well defined. Indeed, let $u' \in [u]$ and $v' \in [v]$. Then by the congruence property

$$u' \cdot v' \in [u \cdot v], \quad \text{i.e.} \quad [u' \cdot v'] = [u \cdot v],$$

so that the operation does not depend on the choice of elements. Furthermore the unit element of the family of sets $U/\sim$ is $[1_U] = \{1_U\}$ since

$$[u] \square [1_U] = [u \cdot 1_U] = [u].$$

Finally

$$[u]^{-1} = [u^{-1}]$$

since

$$[u]\square[u^{-1}]=[u\cdot u^{-1}]=[1_U].$$

In short, we have proved

Theorem 7. The structure $\mathscr{U}/\sim=\langle U/\sim, \square, ^{-1}, [1_U]\rangle$, where $U/\sim$ is the quotient of a set U of units by the equivalence relation $\sim$ of same unit species, is an abelian group written multiplicatively.

We can now justify the customary rule governing unit conversions. The rule is based on

Theorem 8. If $P: A\times B\times\cdots\times N\times U_P\to V$ is a magnitude and $u, u'\in U_P$, then the corresponding values of the associated magnitude P' for u are a linear transform of those for u':

$$P'(a, b, \ldots, n)=p\,u=(\alpha\, p'+\beta)\, u', \qquad \text{with } \alpha, \beta\in R.$$

Example. u = centigrade degree, and u' = Fahrenheit degree, $\alpha=5/9$, $\beta=-5.32/9$. In the vast majority of cases $\beta=0$.

Corollary. 1. If $u, u'\in U_P$ and $\beta=0$, then the units ratio is a unit free real number, namely $u\cdot u^{-1}=p'/p$, called a *conversion factor.*

Corollary 2. The ratio of any two values of an associated magnitude is unit free, hence invariant under unit conversions, provided $\beta=0$.

Units can be freely substituted as long as they are congruent, i.e. provided they belong to the same equivalence class. This substantiates the thesis, occasionally doubted, that *units are conventional*, hence devoid of physical content. (One cause for this doubt is a dislike of conventions for fear that they may lead to conventionalism – a groundless fear no doubt. Another cause is the equally unjustified but widespread confusion of units with dimensions. A third is the confusion of units with standards. More on the latter in Section 2.6.) By the same token, the contention that units are dependent on, or even definable by, measurement is disproved. Rather on the contrary, most measurements presuppose the adoption of units. And, as shown by Axiom 5, such an adoption is an integral part of the choice of representation for physical quantities.

Since units are conventional, we are free to choose them – not however capriciously but in accordance with theory and in the interests of expediency. To the former: (a) fundamental units should be units of physical quantities occurring as fundamental in some basic theory: thus we prefer to choose a fundamental unit of mass rather than weight because, as Gauss noted, mechanics teaches us that mass is more fundamental and universal than weight; (b) all derivative units except the multiples and submultiples of fundamental units are built in accordance with some theoretical formulas. As to the expediency desideratum: although all of the coherent systems of units are rational, not all of them are reasonable or suitable for a given purpose. Some, e.g. the technical system, are cumbersome for theoretical work; others, e.g. the CGS system, are clumsy for practical purposes; finally others, particularly the various system of "natural" units, are devoid of heuristic value.

2.5 Units and Dimensions. Units and dimensions are clearly distinct even though they are often mixed up. Thus, while we can either postulate or compute *the* dimension of a physical quantity, we can only talk of *a* unit (or of a whole class of units) for the same quantity. In other words, whereas the physical quantity-dimension value relation is a function, the physical quantity-unit correspondence is a nonfunctional relation. (Only dimensionless physical quantities are unambiguous as far as units are concerned: they are all assigned the unit 1_U.) In rather obvious symbols: whereas $[P]=d$, where d is a dimension value, $\{P\}=U_P$. We shall presently clarify these ideas by exhibiting their formal relations. To this end we shall avail ourselves of a few results obtained in Section 1, where we introduced a distinction between the dimension function $[\,]$ and the set D_Q of dimension values of a collection Q of physical quantities (magnitudes or dimensional constants). For our immediate purpose we only need to recall that $[\,]$ is a certain function from a set Q of physical quantities to the set D_Q of associated dimension values.

We start off by proving that every dimension value corresponds to a species of units and conversely, as specified by the following

Theorem 9. Let Q be a set of physical quantities and let $\mathscr{D}_Q = \langle D_Q, \cdot, ^{-1}, 1_D\rangle$ be the group of dimension values of Q. Further, let $U_Q/\sim$ be the family of all the species of units for Q. Then D_Q and $U/\sim$ are *isomorphic*, i.e. there exists a bijection

$$\varphi\colon D_Q \to U_Q/\sim .$$

Proof. By Theorem 7 in Section 2.4, $U/\sim$ is an abelian group. By Axiom 3 in Section 1.4, D_Q too is an abelian group. Moreover, for every physical quantity there is just one dimension value and just one species of units, so that the sets D_Q and $U/\sim$ are equally numerous. Finally, the operations are preserved: the products of dimension values are carried into products of units species, and similarly for the inverses. This proves the isomorphism between dimension values and species of units.

Now, every dimension value is in turn a physical quantity species, i.e. a member of the quotient of Q by the relation $\equiv$ of equidimensionality, as shown by Theorem 6 in Section 1.7. Hence the composition of the dimension function $[\,]$ and the above isomorphism φ yields the sought functional relation $\{\,\}$ between physical quantities and species of units, as exhibited by the commutative diagram

$$\begin{array}{ccc} Q & \xrightarrow{[\,]} & D = Q/\equiv \\ & \underset{\{\,\}}{\searrow} & \downarrow \varphi \\ & & U/\sim \end{array}$$

In other words, we have

Theorem 10. Let Q be a set of physical quantities and let $U_Q/\sim$ be the corresponding family of units species. Further, let $[\,]$ be the dimension function that maps Q into the set of dimension values of Q. Then the function $\{\,\}: Q \to U_Q/\sim$ such that $\{q\} = U_q$ for every $q \in Q$, is the composition $\{\,\} = [\,] \circ \varphi$, where φ is the bijection between the dimension values and the units species.

This result elucidates the phrase "a unit of" with the help of the concept of dimension. Note that, on the other hand, the concept of dimension does not depend on the one of unit. This completes our mathematical study of units.

2.6 Units and Standards. Units are sometimes said to be defined by standards. Thus the metre is said to equal, by definition, the length of a certain body laying in state at Sèvres. The corresponding statement is often called a *coordinative definition* following REICHENBACH. But surely this constitutes an abuse of the word "definition" and it fails to elucidate the relations between units and standards.

There are two relations between units and standards, none of which is definitional. Firstly, whenever there happens to be a standard s for a given unit u in the set U_P of P-units, the unit (or rather the denominate number $1 \cdot u$) equals, to within measurement error, the measured value of the associated magnitude P' at the standard s:

$$\text{Measured value of} \quad P'(s) = (1 \pm \sigma)\, u,$$

where σ is the standard deviation of the run of measurements concerned. For a sequence of ideally exact measurements, such as those imagined in the quantum theory of measurements, $\sigma = 0$ and consequently

$$\text{Exact measured value of} \quad P'(s) = P'(s) = 1 \cdot u.$$

Secondly, whenever there is a standard, the corresponding unit may be said to equal, by *referition* (not definition), the P-ness of the given standard:

$$1 \cdot u =_{\mathrm{rf}} P'(s).$$

In either case s is a physical system and more particularly one belonging to the domain of the magnitude in question, whereas $1 \cdot u$ is a concept.

Therefore (a) the unit-standard relation cannot be definitional and (b) the criteria regulating the choice of standards need not be the same as those presiding over the choice of units. For one thing, technological considerations such as feasibility, stability, ease of replication, and cost, which are overriding in the case of standards, are totally irrelevant to the choice of a system of units, which is a theoretical problem. Standards are subject to natural law and to engineering design, and they must be regular and reliable. Units are subject to the algebra of units, they are part of a theoretical framework and they should be convenient. The confusion of units with standards is just an instance of the operationist tenet that every theoretical concept has got to be "defined" by reference to empirical operations (cf. [3]).

2.7 Remarks on the Elucidation Procedure. Units have now ceased to be outcasts: they have been incorporated into the mathematical establishment. Likewise denominate numbers have ceased to be mere "symbolic" products to become charter members of a respectable mathematical structure, namely an R-group. Our characterization of the concept of a unit has not consisted in an explicit definition, let alone in a description of certain measurement procedures: the elucidation has been performed by a theory resting on a couple of postulates (Section 2.2) and containing, among others, two theorems relating units to dimensions (Section 2.5). Consequently the answer to the question "What is a unit?" is as precise as it is sibylline, namely this. A unit is whatever satisfies Axioms 5 and 6 above. Similarly with denominate numbers: a denominate number is an element of the R-group $\mathcal{N}_M^C$. This clarification procedure (the axiomatic definition) is well known from mathematics and there is no other technique for elucidating basic (undefined) concepts, whether in mathematics or in physics (cf. [1] and [4]).

References

1. BUNGE, M.: Foundations of physics. Berlin-Heidelberg-New York: Springer 1967.
2. MACLANE, S., BIRKHOFF, G.: Algebra. New York: Macmillan; London: Collier-Macmillan 1967.
3. BUNGE, M.: Scientific research. Vol. I: The search for system. Vol. II: The search for truth. Berlin-Heidelberg-New York: Springer 1967.
4. – Rev. Modern Phys. **39**, 463 (1967).

2 Reflections on Variational Principles and Invariance Theory

Dominic G. B. Edelen

Center for the Application of Mathematics, Lehigh University
Bethlehem, Pennsylvania

1 Introduction

At this juncture, a minimal requirement for respectability of a physical theory seems to be that it admit a variational principle. This being the case, one eventually becomes aware of a degree of uneasiness, for a variational principle requires the specification of a Lagrangian function and a Lagrangian function is not something that is readily available from knowledge of the physically primitive variables of a phenomena. This follows from the fact that a Lagrangian function for a variational principle must be defined on a space of functions, while the physical primitive variables of a phenomena are a collection of quantities that are quantified by those specific functions that agree with the values of the physical quantities. In general, a Lagrangian function must be defined over a much larger domain of functions than just those that represent the phenomena that we wish to quantify. To be specific, if there is only one variable φ of interest, in addition to the space-time variables, it is usual to consider the Lagrangian function as a function of φ and its first partial derivatives for all functions φ that are C^1 functions on a given domain of the space-time variables. The question thus arises even in this simple problem: how do we determine the Lagrangian function when it can only be sampled for those functions φ that actually arise in physical phenomena. It is obvious that additional assumptions must be made about the phenomena before an identification of the Lagrangian function for a phenomena can be made.

The additional assumptions that must be made in order to obtain a variational principle for a physical phenomena usually fall into two classes. The first class consists of those phenomena for which we possess a system of governing field equations that are adequate in modeling the phenomena. These field equations can then be used to construct a variational principle by the requirement that the governing equations be the Euler equations that arise from the variational principle. This process of imbedding a given system of equations in a variational principle is not a unique process, however: a Lagrangian function that is obtained by this method is only determined to within an equivalence class of such functions, each element of the equivalence class yielding the same Euler equations. There is also a further question that arises with this method. In physical problems we wish to model the whole problem, not just the governing equations; that is, we wish to obtain the governing equations and the admissible classes of initial and boundary data. On the other hand, different types of initial and boundary data require different kinds of variational principles. The additional question thus arises as to what kind of variational principle is to be used.

The second class of assumptions consists of those that require the resultant variational principle to be invariant under the action of a given group of transformations of the space-time variables or of the state variables themselves. Such symmetry requirements are immediately useful in dismissing whole classes of Lagrangian functions as candidates. Most investigations do not use invariance requirements to their fullest extent, however, for there are usually additional assumptions that are made as to the form that the governing equations (resulting Euler equations) must take. The full import of invariance requirements in the determination of Lagrangian functions is thus not readily available. This leaves numerous open questions since we do not know in the general situation to what extent invariance requirements serve to determine Lagrangian functions.

There are many questions, in addition to those we have touched upon above, that will have to be answered before we obtain a full understanding of the significance and the foundations of variational principles in physical theories. We confine ourselves in the space allotted here to certain aspects of the two questions discussed in the previous paragraphs. It is hoped that these reflections may be of use to the current practitioners of the art of construction variational principles. It is further hoped that these reflections may aid in stimulating interest in resolving the numerous and difficult questions that are presented by any approach to variational principles from a foundations point of view.

2 *Preliminary Considerations*

Let D be an open, arcwise simply connected point set in n-dimensional number space X_n, and let D^* denote the closure of D with respect to the Euclidean topology on X_n. The boundary of D^* is denoted by ∂D, and ∂D is assumed to be sufficiently smooth that it forms an $(n-1)$-dimensional hypersurface in X_n such that a normal vector field can be defined at all points of the hypersurface with the exception of a finite number of edges and virtices. The "coordinate differential element" of D^* with respect to the (x)-coordinate cover of D^* is denoted by $dV(x)$, and the "coordinate differential element of directed surface" with respect to the (x)-coordinate cover of D^* is denoted by $\{dS_i(x)\}$. If a coordinate transformation is effected from the (x)-coordinate cover to the (x')-coordinate cover of D^*, the resultant volume element and directed surface element are denoted by $dV(x')$ and $\{dS_{i'}(x')\}$, respectively. All coordinate transformations are assumed to be admissible in the sense that they are one-to-one, of continuity class C^u $(u \geqq 2)$, and globally invertible on an open set of X_n that contains D^* as a proper subset.

Let K denote the class of functions of continuity class C^1 whose domains of definition contain the set D^* and let $\mathscr{D}_1(D^*; N)$ denote the normed linear space whose elements are ordered N-tuples of functions from the class K and whole norm is the uniform convergence norm of the functions and their first partial derivatives. Elements of $\mathscr{D}_1(D^*; N)$ will be denoted by $\{\varphi_\Lambda(x)\}$. The "x" within the parenthesis is used to denote the point P in X_n whose coordinates are the ordered collection of n real numbers $x^1, x^2, \ldots, x^n$ with respect to the (x)-coordinate cover. When it is essential that we distinguish the point from its coordinate labels, we shall write $\{\varphi_\Lambda(P\colon x)\}$ where P denotes the point at which the φ_Λ are to be evaluated and x denotes the coordinate labeling of the point P with respect to the (x)-coordinate cover.

Physical systems are usually described by means of ordered collections of functions that have very particular properties under coordinate transformations; they form vectors, tensors, tensor densities, spinors, components of affine connection, etc. It is therefore useful to introduce the concept of a geometric object field ([1], p. 18), since this concept contains all of the above cases as particularizations.

An ordered collection of N functions on X_n, which has the following properties, is callled a *geometric object of class p* $(p \leqq u)$, or simply a *geometric object* when the value of p is understood:

1. In each coordinate cover of a family of coordinate covers obtained from the (x)-coordinate cover by the class of admissible coordinate transformations of class C^u, it has a well-defined set of N components $\varphi_\Lambda(P;\ )$, in which the capital Greek indices run over the range $1, 2, \ldots, N$.

2. When an admissible coordinate transformation $x^{k'} = f^k(x^m)$ is effected, the new components $\varphi_{\Lambda'}(P;\ )$ at the point $P\colon x(x')$ with respect to the new coordinate cover (x') can be represented as well-defined functions of class C^{u-p} of the old components $\varphi_\Lambda(P;\ )$ at the point $P\colon x$, of the old coordinates x^k, of the functions f^k (of the new coordinates $x^{k'}$), and of the s-th partial derivatives of f^k $(0 \leqq s \leqq p \leqq u)$; that is,

$$\varphi_{\Lambda'}(P;\ ) = F_\Lambda(\varphi_\Sigma(P;\ ), x^k, f^k, \partial_{m_1} f^k, \ldots, \partial_{m_p \ldots m_1} f^k) \tag{2.1}$$

where $\partial_{m_p \dots m_1} f^k \equiv \partial^p f^k / \partial x^{m_p} \partial x^{m_{p-1}} \dots \partial x^m$. We denote the right-hand side of (2.1) by

$$F_\Lambda(\varphi_\Sigma, x^k, x^{k'})$$

for simplicity.

3. The functions $F_\Lambda(\varphi_\Sigma, x^k, x^{k'})$ have the group property:

$$F_\Lambda(F_\Sigma(\varphi_\Gamma, x, x'), x', x'') = F_\Lambda(\varphi_\Sigma, x, x''), \tag{2.2}$$

$$F_\Lambda(F_\Sigma(\varphi_\Gamma, x, x'), x', x) = \varphi_\Lambda, \tag{2.3}$$

$$F_\Lambda(\varphi_\Sigma, x, x) = \varphi_\Lambda. \tag{2.4}$$

We take an arbitrary field $\{\varphi_\Lambda(x)\}$ of a geometric object whose transformation law under coordinate transformations is

$$\varphi_\Lambda = F_\Lambda(\varphi_{\Sigma'}, x', x). \tag{2.5}$$

The law of transformation of $\{\partial_k \varphi_\Lambda\}$ is then given by

$$\partial_k \varphi_\Lambda = F_{k\Lambda}(\varphi_{\Sigma'}, \partial_{k'} \varphi_{\Sigma'}, x', x), \tag{2.6}$$

where the functions $F_{k\Lambda}$ are obtained from the functions F_Λ by partial differentiation with respect to x^k. The quantities $\partial_k \varphi_\Lambda$ are not components of a geometric object since $F_{k\Lambda}$ depends on both $\{\varphi_{\Sigma'}\}$ and $\{\partial_{k'} \varphi_{\Sigma'}\}$. However, $(\varphi_\Lambda, \partial_k \varphi_\Lambda)$ are components of a geometric object whose transformation law is given by

$$\begin{aligned} \varphi_\Lambda &= F_\Lambda(\varphi_{\Sigma'}, x', x), \\ \partial_k \varphi_\Lambda &= F_{k\Lambda}(\varphi_{\Sigma'}, \partial_{k'} \varphi_{\Sigma'}, x', x). \end{aligned} \tag{2.7}$$

It is therefore convenient to consider the composit geometric object $\{{}_\mu\varphi_\Lambda\}$ $(\mu = 0, 1, \dots, n)$ defined by

$${}_0\varphi_\Lambda = \varphi_\Lambda, \qquad {}_k\varphi_\Lambda = \partial_k \varphi_\Lambda, \tag{2.8}$$

since we shall require both the geometric object field and its first partial derivatives in the variational theory.

Let $L = L(x^k, {}_\mu\varphi_\Lambda(x^k))$ be a given function of class C^2 in all of its $n + N(n+1)$ arguments. With L as the Lagrangian, we than have the continuous and differentiable functional

$$J[\{\varphi_\Lambda\}] = \int_{D^*} L(x^k, {}_\mu\varphi_\Lambda(x^k))\, dV(x) \tag{2.9}$$

with domain $\mathscr{D}_1(D^*; N)$. Let $\{\varphi_{\Lambda 0}(x)\}$ designate a given element of $\mathscr{D}_1(D^*; N)$ and set

$$\tilde{\varphi}_{\Lambda 0}(x) = \varphi_{\Lambda 0}(x) + h_\Lambda(x), \tag{2.10}$$

where $\{h_A(x)\} \in \mathscr{D}_1(D^*; N)$; we refer to $\{h_A(x)\}$ as the *variation* of $\{\varphi_{A0}(x)\}$. An element $\{\varphi_{A0}(x)\}$ of $\mathscr{D}_1(D^*; N)$ is said to render $J[\{\varphi_A\}]$ *stationary* (in the classical sense) if and only if

$$J[\{\tilde{\varphi}_{A0}\}] = J[\{\varphi_{A0}\}] + o(\|\{h_A\}\|) \tag{2.11}$$

holds for all variations $\{h_A(x)\}$ that vanish on ∂D^*. An element $\{\varphi_{A0}(x)\}$ is said to render $J[\{\varphi_A\}]$ *stationary relative to given boundary data* if and only if (2.11) holds for all variations $\{h_A(x)\}$ when $\{\varphi_{A0}\}$ satisfies the given boundary data. Let $(\partial D^*)_1$ and $(\partial D^*)_2$ be a decomposition of ∂D^* such that $\partial D^* = (\partial D^*)_1 \cup (\partial D^*)_2$ and such that $(\partial D^*)_1$ and $(\partial D^*)_2$ each consist of finitely many arcwise simply connected manifolds of dimension $(n-1)$ in X_n. An element $\{\varphi_{A0}(x)\}$ of $\mathscr{D}_1(D^*; N)$ is said to render $J[\{\varphi_A\}]$ *stationary relative to given boundary data on* $(\partial D^*)_1$ if and only if (2.11) holds for all variations $\{h_A(x)\}$ that vanish on $(\partial D^*)_2$ when $\{\varphi_{A0}(x)\}$ satisfies the given boundary data on $(\partial D^*)_1$.

The process of rendering $J[\{\varphi_A\}]$ stationary is the natural process when we are given the values of $\{\varphi_{A0}\}$ on ∂D^*, for the requirement that the variations vanish on ∂D^* simply insures that both $\{\varphi_{A0}\}$ and $\{\tilde{\varphi}_{A0}\}$ satisfy the given data for all variations. The process of rendering $J[\{\varphi_A\}]$ stationary relative to given data is the natural process when $\{\varphi_{A0}\}$ is required to satisfy conditions on its directional derivative at the boundary; while rendering $J[\{\varphi_A\}]$ stationary relative to given boundary data on $(\partial D^*)_1$ is the natural process when the values of $\{\varphi_{A0}\}$ are given on $(\partial D^*)_2$ while $\{\varphi_{A0}\}$ is required to satisfy conditions on its directional derivative on $(\partial D^*)_1$. These three kinds of stationarization processes loosely correspond to *stationarization problems* (EULER equations plus given data) of the DIRICHLET, NEUMANN, and mixed types, respectively. Lumping these three kinds of stationarization processes together can lead to conceptual difficulties and to incorrect conclusions.

Since the Lagrangian function, L, is assumed to be of class C^2 in its arguments, it is an easy matter to show that

$$\begin{aligned} J[\{\tilde{\varphi}_A\}] - J[\{\varphi_A\}] &= \int_{D^*} \{E|L\}_{\varphi_A}(x^k)\, h_A(x^k)\, dV(x) + o(\|\{h_A\}\|) \\ &\quad + \int_{\partial D^*} \frac{\partial L(x^k, {}_\mu\varphi_\Sigma(x^k))}{\partial(\partial_m \varphi_A(x^k))} N_m(x^k)\, h_A(x^k)\, dS(x^k), \end{aligned} \tag{2.12}$$

where

$$\{E|L\}_{\varphi_A}(x^k) = \frac{\partial L}{\partial \varphi_A} - \partial_m \left(\frac{\partial L}{\partial(\partial_m \varphi_A)}\right) \tag{2.13}$$

is the Euler-Lagrange operator with respect to φ_A and we have made the substitution $dS_m(x) = N_m(x)\, dS(x)$ with $\{N_m(x)\}$ as the unit normal vector field of ∂D^* (outwardly oriented) and $dS(x)$ is the scalar differential surface measure on ∂D^*.[1] Standard arguments based upon the classic lemma of the calculus of variations give the following conclusions if the conditions for C^2 solutions of the EULER equations are met (see previous footnote). *The functions* $\varphi_{A0}(x)$ *render* $J[\{\varphi_A\}]$ *stationary if and only if the Euler equations*

$$\{E|L\}_{\varphi_{A0}}(x^k)=0, \qquad A=1,\ldots,N \tag{2.14}$$

are satisfied at all points in **D**. *The functions* $\varphi_{A0}(x)$ *render* $J[\{\varphi_A\}]$ *stationary relative to given boundary data if and only if the Euler equations* (2.14) *are satisfied at all points in* **D** *and the given boundary data is such that*

$$\frac{\partial L}{\partial(\partial_m \varphi_A)} N_m=0, \qquad A=1,\ldots,N \tag{2.15}$$

hold at all points on ∂D^*. *The functions* $\varphi_{A0}(x)$ *render* $J[\{\varphi_A\}]$ *stationary relative to given boundary data on* $(\partial D^*)_1$ *if and only if the Euler equations* (2.14) *are satisfied at all points in* **D** *and the given boundary data satisfies the conditions* (2.15) *at all points on* $(\partial D^*)_1$. The reader should note that when there is no specification about boundary data on ∂D^* or on a part of ∂D^*, then we can require $\{\varphi_{A0}\}$ to assume given values on ∂D provided such specifications are compatable with the EULER equations. We thus have the standard results of the calculus of stationarization of functionals, as well as precise conditions wherein "natural boundary data" problems are couched.

An important class of Langrangian functions are those for which

$$\{E|A\}_{\varphi_A}(x^k)=0, \qquad A=1,\ldots,N \tag{2.16}$$

hold for every element $\{\varphi_A\}$ of $\mathscr{D}_1(D^*;N)$. This class is referred to as the *null class*, $\mathscr{N}(N)$, *of the Euler-Lagrange operator* and has the property that any Lagrangian function A belonging to $\mathscr{N}(N)$ is such that every collection of functions $\{\varphi_A\}$ belonging to $\mathscr{D}_1(D^*;N)$ renders $J[\{\varphi_A\}]$ stationary in value for all variations that vanish on the boundary of D^*. Thus, if $A\in\mathscr{N}(N)$ and if L is a Lagrangian function such that a given system of equations are the EULER equations $\{E|L\}_{\varphi_A}(x^k)=0$, then we also have

$$\{E|L+A\}_{\varphi_A}(x^k)=0. \tag{2.17}$$

This simply says that the Lagrangian function that produces a given system of equations as EULER equations is only determined to within

[1] It is obvious that in making use of the EULER-LAGRANGE operator in (2.12) we have assumed that the functions that occur in the Lagrangian function will turn out to be of class C^2, even though the functionals under study have $\mathscr{D}_1(D^*;N)$ as domain. In almost all problems of physical interest the conditions that give C^2 solutions to the EULER equations are satisfied (the HILBERT differentiability theorem). The reader interested in this aspect of the problem is referred to [2] for a full discussion.

an equivalence class $\mathscr{E}(L)$ defined by

$$B \in \mathscr{E}(L) \Leftrightarrow B = L + A \quad \text{for some } A \in \mathscr{N}(N).$$

However, even though the EULER equations are unchanged under addition of an element of the null class of the EULER-LAGRANGE operator, this is *not* true in general for the boundary data conditions (2.15):

$$\frac{\partial(L+A)}{\partial(\partial_m \varphi_A)} N_m = \frac{\partial L}{\partial(\partial_m \varphi_A)} N_m + \frac{\partial A}{\partial(\partial_m \varphi_A)} N_m,$$

and hence

$$\frac{\partial(L+A)}{\partial(\partial_m \varphi_A)} N_m = \frac{\partial L}{\partial(\partial_m \varphi_A)} N_m = 0$$

only if A is such that

$$\frac{\partial A}{\partial(\partial_m \varphi_A)} N_m = 0 \tag{2.18}$$

holds on ∂D^* or on the appropriate part of ∂D^*. The detailed characterization of the null class of the EULER-LAGRANGE operator is thus of great importance in any problem of imbedding systems of equations as variational statements or for variational problems relative to given boundary data. The reader is referred to [3], Section 6; [4], Section 11; [5], Section 1.6 and 2.3 for the details of the characterization. It turns out that if A depends on $\{\varphi_A\}$ then A must have the form of a divergence of a vector valued function of $(\varphi_A(x^k), x^k)$, and hence $\partial A/\partial(\partial_m \varphi_A)$ will not vanish and will not satisfy (2.18), in general. This result should not come as a surprise to the reader, for invariance of the EULER equations does not necessarily have anything to do with invariance of a class of problems; a problem consisting of EULER equations plus boundary data. The result given above for the case of just stationarization does lead to invariance of the problem when the EULER equations are left invariant. On the other hand, whenever a problem involves boundary data that places constraints on the directional derivative at the boundary, then invariance of the EULER equations does not imply invariance of the problem. This is also an essential point in invariance theory, for invariance of the EULER equations under coordinate or point transformations is sufficient for invariance of the generating functional of the EULER equations only if the boundary data does not involve derivatives of the unknown functions. Stated another way, invariance of the problems (differential equations plus given boundary date) is what is required, not just the invariance of the differential equations. Those authors who say that the image of a functional under coordinate or point transformations must agree with the original functional to within an element of the null class of the EULER-LAGRANGE operator, obtain invariance only for the EULER equations, not for problems. To put this observation in a concrete context, we start with the functional

$$J[\{\varphi_A\}] = \int_{D^*} L(x^k, \varphi_A(x^k), \partial_m \varphi_A(x^k))\, dV(x) \tag{2.19}$$

and then form the functional

$$\begin{aligned}\tilde{J}[\{\varphi_A\}] &= \int_{D^*} L(x^{k'}, \varphi_{A'}(x^{k'}), \partial_{m'} \varphi_{A'}(x^{k'}))\, dV(x') \\ &= \int_{D^*} L(f^k(x^t), F_A(\varphi_\Sigma, x^k, f^k), F_{mA}(\varphi_\Sigma, \partial_t \varphi_\Sigma, x^k, f^k)) \\ &\quad \cdot \left| \frac{\partial(f^k)}{\partial(x^k)} \right| dV(x) \\ &= \int_{D^*} L'(x^k, \varphi_A(x^k), \partial_m \varphi_A(x^k))\, dV(x)\end{aligned} \tag{2.20}$$

for a class of admissible coordinate transformations $x^{k'} = f^k(x^m)$. Invariance of the EULER equations is guaranteed by the requirement

$$L' = L + A \tag{2.21}$$

for some A belonging to $\mathcal{N}(N)$. Here L' is given as a function of $(x^k, {}_\mu\varphi_A(x^k))$ by (2.20). Such a procedure will only lead to invariance for problems of the DIRICHLET type. Problems of the NEUMANN or mixed types will not be invariant since the boundary data will change if A depends on $\{\varphi_A\}$. If, indeed, all problems of physical interest are problems of DIRICHLET type, then all is well and good.

3 Extended Functional Variations

Consider the r-parameter family $T(\boldsymbol{\lambda})$ of admissible coordinate transformations

$$x^{k'} = G^k(x^m; \lambda^\alpha) \tag{3.1}$$

of class C^u in x^m for all values of the parameters $\{\lambda^\alpha\}$ $(\alpha = 1, \ldots, r)$ in some open set of E_r that contains the identity element of addition in E_r and such that

$$G^k(x^m; \lambda^\alpha) = x^k + \lambda^\alpha v^k_\alpha(x^m) + o(\|\boldsymbol{\lambda}\|) \tag{3.2}$$

where the r-tuple of vectors $v^k_\alpha(x^m)$ $(\alpha = 1, \ldots, r)$ are the infinitesimal generators of the transformations $T(\lambda)$.

If $\{\varphi_A(x)\}$ is a geometric object, then in a sufficiently small neighborhood of $\boldsymbol{\lambda} = \mathbf{0}$, the equations of transformation for such quantities give (see [6], pp. 107–8 and 112–3)

$$\varphi_{A'}(x) = \varphi_A(x) + \lambda^\alpha \underset{v_\alpha}{D}\, \varphi_A(x) + o(\|\boldsymbol{\lambda}\|), \tag{3.3}$$

where

$$\underset{v_\alpha}{D}\,\varphi_A(x) = v_\alpha^k\,\partial_k\,\varphi_A(x) - \underset{v_\alpha}{£}\,\varphi_A(x) \tag{3.4}$$

and £ denotes the process of Lie differentiation [1]. In view of (3.3), we refer to the quantities $\{\underset{v_\alpha}{D}\,\varphi_A\}$ as the coordinate variations of $\{\varphi_A\}$ that are induced by the transformations $T(d\boldsymbol{\lambda})$. A rudimentary calculation based on (3.3) gives us

$$\frac{\partial\varphi_{A'}}{\partial x^{k'}}(x) = \partial_k\,\varphi_A(x) + o(\|\boldsymbol{\lambda}\|) - \lambda^\alpha\big[\partial_k\,v_\alpha^m\,\partial_m\varphi_A - \partial_k(\underset{v_\alpha}{D}\,\varphi_A)\big]\,(x). \tag{3.5}$$

It thus follows that all of the arguments of Lagrangian functions considered here have well defined values under the action of coordinate transformations. The question of point transformations is easily handled on the basis of the above formalism if we adjoin suitable function variations of the φ_A, see [5, 7, 8]. We may accordingly form the functionals

$$\tilde{J}[\{\varphi_A\}] = \int_{D^*} L(x^{k'}, {}_{\mu'}\varphi_{A'}(x^{k'}))\,dV(x'). \tag{3.6}$$

The difference

$$\Delta J[\{\varphi_A\}] = \tilde{J}[\{\varphi_A\}] - J[\{\varphi\}] \tag{3.7}$$

is referred to as the *extended finite variation of J* that is induced by coordinate transformations. Written out in full, (3.7) gives

$$\Delta J[\{\varphi_A\}] = \int_{D^*} L(x^{k'}, {}_{\mu'}\varphi_{A'}(x^{k'}))\,dV(x') - \int_{D^*} L(x^k, {}_{\mu}\varphi_A(x^k)\,dV(x).$$

The *infinitesimal extended variation* of $J[\{\varphi_A\}]$ is obtained from the finite extended variation of $J[\{\varphi_A\}]$ by series expansion in the parameters $\boldsymbol{\lambda}$ about $\boldsymbol{\lambda} = \mathbf{0}$:

$$\begin{aligned}\Delta J[\{\varphi_A\}] = \int_{D^*} L\big(&x^k + \lambda^\alpha v_\alpha^k + o(\boldsymbol{\lambda}),\ \varphi_A(x) + \lambda^\alpha \underset{v_\alpha}{D}\,\varphi_A(x) + o(\boldsymbol{\lambda}),\\ &\partial_k\,\varphi_A(x) + \lambda^\alpha[\partial_k\,v_\alpha^m\,\partial_m\,\varphi_A\\ &\quad - \partial_k(\underset{v_\alpha}{D}\,\varphi_A)]\,(x) + o(\lambda)\big)\,dV(x + \lambda^\alpha v_\alpha + o(\lambda))\\ &- \int_{D^*} L(x^k, \varphi_A(x), \partial_k\,\varphi_A(x))\,dV(x).\end{aligned} \tag{3.8}$$

Since L is assumed to be of class C^2 in its arguments, the series expansion of (3.8) is equivalent to obtaining the linear functional approximation of the functional $\tilde{J}$ that results when $\tilde{J}$ is considered as a function of the parameters λ. Thus, if we write

$$\Delta J[\{\varphi_\lambda\}] = \lambda^\alpha \int_{D^*} J_\alpha\,dV(x) + o(\|\lambda\|), \tag{3.9}$$

a straightforward calculation yields

$$\begin{aligned} J_\alpha = & L\partial_m v_\alpha^m + v_\alpha^m \partial_m^* L + \frac{\partial L}{\partial \varphi_A} \underset{v_\alpha}{D} \varphi_A \\ & + \frac{\partial L}{\partial(\partial_k \varphi_A)} \{\partial_k(\underset{v_\alpha}{D} \varphi_A) - \partial_k v_\alpha^m \partial_m \varphi_A\}, \end{aligned} \tag{3.10}$$

where ∂_k^* is defined by $\partial_k L = \partial_k^* L + \frac{\partial L}{\partial({}_\mu\varphi_A)} \partial_k({}_\mu\varphi_A)$. There are several equivalent forms of J_α which we shall now obtain. If we use the relations

$$\underset{v_\alpha}{D} \varphi_A = v_\alpha^k \partial_k \varphi_A - \underset{v_\alpha}{£} \varphi_A \tag{3.11}$$

and collect the various terms together, we have

$$J_\alpha = \partial_m(L v_\alpha^m) - \frac{\partial L}{\partial \varphi_A} \underset{v_\alpha}{£} \varphi_A - \frac{\partial L}{\partial(\partial_k \varphi_A)} \partial_k(\underset{v_\alpha}{£} \varphi_A). \tag{3.12}$$

If we introduce the EULER-LAGRANGE operators, we have

$$J_\alpha = \partial_k \left(L v_\alpha^k - \frac{\partial L}{\partial(\partial_k \varphi_A)} \underset{v_\alpha}{£} \varphi_A\right) - \{E|L\}_{\varphi_A}(x) \underset{v_\alpha}{£} \varphi_A, \tag{3.13}$$

$$J_\alpha = \partial_k \left(\frac{\partial L}{\partial(\partial_k \varphi_A)} \underset{v_\alpha}{D} \varphi_A - T_m^k v_\alpha^m\right) - \{E|L\}_{\varphi_A}(x) \underset{v_\alpha}{£} \varphi_A, \tag{3.14}$$

where the last form comes using (3.11) in (3.13) and T_m^k denotes the momentum energy complex formed from L:

$$T_m^k = \frac{\partial L}{\partial(\partial_k \varphi_A)} \partial_m \varphi_A - \delta_m^k L. \tag{3.15}$$

This collection of functions of $\{{}_\mu\varphi_A\}$ satisfy the identities

$$\begin{aligned} \partial_k T_m^k & \equiv -\{E|L\}_{\varphi_A}(x) \partial_m \varphi_A - \partial_m^* L, \\ \partial_m L & = \partial_m^* L + \frac{\partial L}{\partial({}_\mu\varphi_A)} \partial_m({}_\mu\varphi_A). \end{aligned} \tag{3.16}$$

Another useful form can be obtained by using (3.15) to simplify (3.10):

$$J_\alpha = v_\alpha^k \partial_k^* L + (\underset{v_\alpha}{D} \varphi_A) \frac{\partial L}{\partial \varphi_A} + (\partial_m \underset{v_\alpha}{D} \varphi_A) \frac{\partial L}{\partial(\partial_m \varphi_A)} - T_m^k \partial_k v_\alpha^m. \tag{3.17}$$

The forms given by (3.12), (3.13), (3.14), and (3.17) form the basis upon which one erects invariance theory. There are many aspects of invariance theory that are almost virgin territory and which bear significant additional work, even though invariance theory is of such importance in physical theory. The limitation space in this paper preclude our investigation of more that one of these aspects here. The reader is referred to [5] for statements of some of the unresolved problems and for a listing of a number of different kinds of invariance requirements that can be associated with functionals.

4 *Invariance as a Sieve for Lagrangians*

For any n-dimensional subset d of D^* and for any Lagrangian L of class C^2 in its arguments, we may form the functional

$$J[\{\varphi_A\}](d)=\int_{d^*} L(x^k, {}_\mu\varphi_A(x^k))\, dV(x) \tag{4.1}$$

on the space $\mathscr{D}_1(D^*; N)$. If we introduce an r-parameter family of admissible coordinate transformations we can then form

$$\tilde{J}[\{\varphi_A\}](d)=\int_{d^*} L(x^{k'}, {}_{\mu'}\varphi_{A'}(x^{k'}))\, dV(x'), \tag{4.2}$$

and obtain

$$\Delta J[\{\varphi_A\}](d)=\lambda^\alpha \int_{d^*} J_\alpha\, dV(x)+o(\|\lambda\|) \tag{4.3}$$

in exactly the same way as that used in the previous section. Now, we have used the same function L in forming both J and $\tilde{J}$. Hence, if the values of J and $\tilde{J}$ are different for any $\{\varphi_A\}\in\mathscr{D}_1(D^*; N)$, this difference arises because we have used different coordinate covers on D^*. The study of the conditions that insure that $J[\{\varphi_A\}]$ will be unchanged under admissible changes in coordinate covers of D^* is the *direct problem of invariance.* The class of problems that we wish to consider here may be referred to as *indirect problems of invariance.* These problems are characterized as a search for Lagrangian functions such that $J[\{\varphi_A\}]$ formed from these Lagrangian functions will be unchanged when acted upon by coordinate transformations that constitute a given finite parameter continuous group of coordinate transformations. In this way we shall be able to construct physical theories when we know only that the governing equations and boundary data of the theory must be invariant under the action of a given group. Such a procedure will not involve the difficult problem of attempting to imbed a given system of equations as EULER equations nor will it involve attempts to infer the Lagrangian function from only a finite number of tested solutions of EULER equations.

If an r-parameter continuous group G_r of coordinate transformations leaves $J[\{\varphi_A\}](d)$ unchanged in value for all $d\subset D$ when G_r is restricted to a sufficiently small neighborhood of the identity element, then the Lagrangian function, L, that serves to define $J[\{\varphi_A\}](d)$ is said to admit G_r as a local coordinate invariance group. The collection of all Lagrangian functions *that admits G_r as a local coordinate invariance group is referred to as the Lagrangian coordinate class $L_c(G_r)$.*

With the above definition, our search for Lagrangian functions for which $J[\{\varphi_A\}]$ is unchanged under the action of a *given* r-parameter continuous group G_r is equivalent to the problem of characterizing the collection $L_c(G_r)$. When (3.7) and (3.9) are used together with the fundamental lemma of the calculus of variations, it is easily seen that $J[\{\varphi_A\}](d)$ is unchanged in value for every $d\subset D$ if and only if

$$J_\alpha=0, \qquad \alpha=1,\ldots,r, \tag{4.4}$$

where the functions $\{v_\alpha^k(x)\}$, $\alpha=1,\ldots,r$ are the infitinesimal generating vectors of the group G_r. Accordingly, when (3.10) is used, we obtain the following conclusion.

Necessary and sufficient conditions that $L\in L_c(G_r)$, where G_r is a given r-parameter continuous group with infinitesimal generating vectors $\{v_\alpha^k(x)\}$, are that L satisfies the system of r first order partial differential equations

$$\begin{aligned} 0=L\,\partial_m v_\alpha^m+v_\alpha^m\,\partial_m^* L+\big(\underset{v_\alpha}{D}\,\varphi_A\big)\frac{\partial L}{\partial\varphi_A}\qquad&\\ +\big(\partial_k(\underset{v_\alpha}{D}\,\varphi_A)-\partial_k v_\alpha^m\,\partial_m\varphi_A\big)\frac{\partial L}{\partial(\partial_k\varphi_A)},\qquad&\alpha=1,\ldots,r. \end{aligned} \tag{4.5}$$

If $L\in L_c(G_r)$, we immediately obtain the various strong and weak conservation laws and laws of balance from the equivalence of (3.10), (3.12), (3.13), and (3.14). For example, *if $L\in L_c(G_r)$, then we have the r identities*

$$\partial_k\left\{\frac{\partial L}{\partial(\partial_k\varphi_A)}\underset{v_\alpha}{D}\,\varphi_A-T_m^k\,v_\alpha^m\right\}=\{E\,|\,L\}_{\varphi_A}(x)\underset{v_\alpha}{\pounds}\,\varphi_A \tag{4.6}$$

where the r vector fields $\{v_\alpha^k(x)\}$ are the r infinitesimal generating vectors of G_r.

The above results tell us that $L\in L_c(G_r)$ if and only if L satisfies the system of r first order partial differential equations (4.5). Now, by assumption, L is a function of the $n+N(n+1)$ arguments

$$x^k,\ \varphi_A,\ \partial_m\varphi_A\equiv P_{mA}. \tag{4.7}$$

When (3.4) is used, it is easily seen that $\underset{v_\alpha}{D}\,\varphi_A$ *and* $\partial_k(\underset{v_\alpha}{D}\,\varphi_A)$ are functions of only the $n+N(n+1)$ arguments given in the list (4.7). It thus follows that the system (4.5) constitutes a system of r quasilinear first order partial differential equations for the determination of a functions of the arguments given in the list (4.7). The classic theory of such systems is thus directly applicable. This theory tells us that if γ denotes the number of independent equations in the completion of the system (4.5), then the general solution of (4.5) is an arbitrary C^2 (since L is required to be of class C^2) function of the $n+N(n+1)-\gamma$ primitive integrals of the completion of (4.5). Thus, the search for Lagrangian functions by the requirement that $L \in L_c(G_r)$ for a given G_r has the added advantage of reducing the number of arguments that enter into the specification of such an L by exactly the number of independent equations in the completion of the system (4.5). Thus, if the system (4.5) is complete and independent, the number of arguments that serve to determine L is $n+N(n+1)-r$. This procedure also has an additional advantage. The only assumptions that have been made in determining $L_c(G_r)$ is that all problems associated with $J[\{\varphi_A\}]$ formed from any $L \in L_c(G_r)$ are unchanged under the action of G_r in a sufficiently small neighborhood of the identity element of G_r. Determining $L_c(G_r)$ by obtaining the general solution of (4.5) thus determines the class of *all* Lagrangian functions with this property. Theories based upon such a procedure are thus universal inveloping theories for all possible theories that admit G_r as a coordinate invariance group. We haven't really obtained something for nothing, however, for the construction of the system (4.5) demands that we compute $\{\underset{v_\alpha}{D}\varphi_A\}$ and this in turn demands that we must assume the transformation properties of the N-tuple $\{\varphi_A\}$ before we start; you have to know whether the Lagrangian depends on scalar fields, or vector fields, or tensor fields or what have you, before you start.

There are obviously many interesting problems that come to mind at this point that must be foregone in a paper of this length (see [5]). We simple conclude with three simple examples of the method.

Example I. $n=2$, $N=1$, $L=L(x^k, \varphi_1, \partial_m \varphi_1)$, $\varphi_1(x)=$ scalar function so that $\underset{v_\alpha}{D}\varphi_1=0$, $r=3$, $G_3=$3-parameter Lorentz group on X_2,

$$G_3: \begin{cases} v_1^1=1, & v_2^1=0, \quad v_3^1=x^2 \\ v_1^2=0, & v_2^2=1, \quad v_3^2=a^2 x^1. \end{cases}$$

The system (4.5) is the complete system

$$\alpha=1, \quad \frac{\partial^* L}{\partial x^1}=0; \qquad \alpha=2, \quad \frac{\partial^* L}{\partial x^2}=0;$$

$$\alpha=3, \quad x^2 \frac{\partial^* L}{\partial x^1}+a^2 x^1 \frac{\partial^* L}{\partial x^2}-\partial_1 \varphi_1 \frac{\partial L}{\partial(\partial_2 \varphi_1)}-a^2 \partial_2 \varphi_1 \frac{\partial L}{\partial(\partial_1 \varphi_1)}=0.$$

The general solution is given by

$$L=L\big(\varphi_1, (\partial_1 \varphi_1)^2-a^2(\partial_2 \varphi_1)^2\big)$$

and indeed $n+N(n+1)-r=2+1(2+1)-3=2$.

Example II. $n=2$, $N=1$, $L=L(x^k, \varphi_1, \partial_m \varphi_1)$, $\varphi_1(x)=$scalar function so that $\underset{v_\alpha}{D} \varphi_1=0$, $r=3$, $G_3=3$-parameter Euclidean group on X_2,

$$G_3: \begin{cases} v_1^1=1, & v_2^1=0, & v_3^1=x^2 \\ v_1^2=0, & v_2^2=1, & v_3^2=-x^1. \end{cases}$$

The system (4.5) is the complete system

$$\alpha=1, \quad \frac{\partial^* L}{\partial x^1}=0; \quad \alpha=2, \quad \frac{\partial^* L}{\partial x^2}=0;$$

$$\alpha=3, \quad x^2 \frac{\partial^* L}{\partial x^1} - x^1 \frac{\partial^* L}{\partial x^2} - \partial_1 \varphi_1 \frac{\partial L}{\partial(\partial_2 \varphi_1)} + \partial_2 \varphi_1 \frac{\partial L}{\partial(\partial_1 \varphi_1)}=0.$$

The general solution is given by

$$L=L\big(\varphi_1, (\partial_1 \varphi_1)^2+(\partial_2 \varphi_1)^2\big).$$

Example III. This is the same as example I with G_3 replaced by

$$G_1: \begin{cases} v_1^1=x^2 \\ v_1^2=a^2 x^1 \end{cases}$$

and $r=1$. The system (4.5) becomes the single equation

$$x^2 \frac{\partial^* L}{\partial x^1} + a^2 x^1 \frac{\partial^* L}{\partial x^2} - \partial_1 \varphi_1 \frac{\partial L}{\partial(\partial_2 \varphi_1)} - a^2 \partial_2 \varphi_1 \frac{\partial L}{\partial(\partial_1 \varphi_1)}=0.$$

The general solution is given by

$$L=L\big(a^2 (x^1)^2-(x^2)^2, \varphi_1, (\partial_1 \varphi_1)^2-a^2(\partial_2 \varphi_1)^2, a^2 x^1 \partial_2 \varphi_1 + x^2 \partial_1 \varphi_1\big).$$

References

1. YANO, K.: The theory of Lie derivatives and its applications. Amsterdam: North-Holland 1957.
2. HESTENES, M. E.: Calculus of variations and optimal control theory. New York: John Wiley 1966.
3. EDELEN, D. G. B.: Internat. J. Engrg. Sci. **7**, 373–389 (1969).
4. – Internat. J. Engrg. Sci. **7**, 391–399 (1969).
5. – Nonlocal variations and local invariance of fields. New York: American Elsevier 1969.
6. SCHOUTEN, J. A.: Ricci calculus (2nd edit.). Berlin-Göttingen-Heidelberg: Springer 1954.
7. EDELEN, D. G. B.: Nuovo Cimento **25**, 778–792 (1962).
8. – Nuovo Cimento **25**, 793–805 (1962).

II. Of Particles

3 Galilei- and Lorentz-Invariant Particle Systems and their Conservation Laws[1]

Peter Havas

Department of Physics

Temple University, Philadelphia

1 *Introduction*

Newton's mechanics of point particles has served as a model of physical theory for two centuries. In its usual formulation, the forces considered are two-body forces depending on the mutual separation of the particles at the same time. The third law implies that these forces are derivable from a potential also depending only on this separation, which allows a Lagrangean formulation of the theory. As was gradually recognized during the past century, this particular form of Newtonian mechanics is invariant under a ten-parameter group, the inhomogeneous Galilei group [1], and possesses ten integrals which are algebraic functions of the positions and velocities; one of these corresponds to the conservation of energy, three to the conservation of linear momentum, three to the conservation of angular momentum, and three express the uniform motion of the center of mass.

These integrals can be readily obtained by direct integration of Newton's equations of motion [2]. However, new aspects of their significance emerge through their derivation, on the basis of Noether's theorem [3, 4], from the invariance properties of the Lagrangean under the infinitesimal transformations of the inhomogeneous Galilei group [5, 4]. The integrals are related respectively to the time translation, space translations, space rotations, and the Galilei transformations between inertial systems in relative motions.

By the time these connections had been fully realized, the concepts of Newtonian physics had been superseded by those of the special theory of relativity. Concurrently, the idea that particles and their interactions were the basis of all physical phenomena had gradually been replaced by the belief that this basis was provided by fields, especially the electromagnetic and the gravitational one [6]. Indeed, it had been widely taken

[1] Research supported in part by the National Science Foundation.

for granted that the new features of the concept of causality introduced by the theory of relativity *required* a description by means of fields rather than by direct particle interactions, and little work was done in developing the theory of such interactions. Only in the last two decades has it been widely realized that a relativistic theory of interacting particles is not only possible, but may be of help in solving some of the difficulties of present-day theory [7, 8]; furthermore, after decades of stressing the differences, it was recognized by several authors that Newtonian and relativistic theories possess many fundamental features which are remarkably similar [9].

The fundamental symmetry group of the special theory of relativity is the Lorentz group. It is a 10-parameter group, just like the Galilei group. If the equations of motion of a special relativistic system of point particles can be derived from a "Fokker-type" variational principle [10–12] describing direct particle interactions, this system too possesses ten integrals which are constants of motion. However, the forces acting in such a system do not depend on the simultaneous positions of the particles only; thus the constants of the motion do not depend on the position and velocities of the particles at any single time, in sharp contrast to the Newtonian conserved quantities, but involve integrals extended over the world lines of the particles. It has been shown that (except for the trivial case of no interaction) the constants of motion of any special relativistic dynamics must be of this type [13].

Possible generalizations of Newtonian mechanics allowing for more general forces, but maintaining the conservation laws, were considered even before the connection of these laws with invariance properties of the equations was fully realized. From this point of view generalized translation-invariant forces seem to have been discussed first by SCHÜTZ [14]; derivatives of arbitrary order in the Lagrangean were considered by KÖNIGSBERGER [15]. In these generalizations, the forces were still assumed to depend on the particle variables taken at the same instant of time. This restriction was also maintained in recent papers which took invariance under the Galilei group as their starting point for the investigation of possible forms of two-body [16] and n-body interactions [17]. It was realized only recently [18] that Galilei-invariance alone does not lead to such a restriction, but allows for forces which depend on the particle variables at different times, and thus permits a mechanics of particles very similar to the special relativistic one, with conservation laws of the same type.

Section 2 contains a brief review of the Newtonian conservation laws and their connection with NOETHER's theorem. Variational principles for systems of particles with general Galilei- and Lorentz-invariant two-body interactions are presented in Section 3; the equations of motion following from these principles are obtained, and the possibility of adjunct field theories is discussed. The conservation laws of such systems following from Galilei or Lorentz invariance are found in Section 4, as well as an additional conservation law valid for a class of interactions invariant under scale transformations; the question of an alternative formulation of the conservation laws is discussed, and the possibility of generalizing the results obtained is indicated.

2 *Noether's Theorem and Newtonian Conservation Laws*

The fundamental connection between the invariance properties of a variational principle and the conservation laws has been expressed in two theorems by E. NOETHER [3, 19], referring to the invariance under a finite and an infinite continuous group, respectively. In the present context we are only concerned with the first case of a group G_P depending on p parameters. Furthermore, we shall restrict ourselves in this section to the consideration of an integral

$$I=\int L(x^i, v^i, \tau)\, d\tau, \qquad i=1\dots N,\; v^i\equiv\frac{dx^i}{d\tau}, \tag{1}$$

extended over a one-dimensional region only, and not containing higher derivatives of the functions x^i than the first with respect to the arbitrary parameter τ. As usual we introduce the Lagrangean derivatives

$$L_i\equiv\frac{\partial L}{\partial x^i}-\frac{d}{d\tau}\frac{\partial L}{\partial v^i}. \tag{2}$$

We now consider the infinitesimal transformations

$$\tau'=\tau+\delta\tau, \qquad x^{i\prime}=x^i+\delta x^i, \tag{3}$$

where $\delta\tau$, δx^i are functions of t, x^i, v^i depending linearly on the infinitesimal parameters of G_p. Then NOETHER's first theorem states: *If I is invariant under the infinitesimal transformations of G_p up to a divergence, there exist precisely p linearly independent combinations of the Lagrangean derivatives which are divergences. Conversely, if p such combinations are divergences, there exists a set of p linearly independent infinitesimal transformations which leave I invariant up to a divergence. These transformations generate a G_p provided the x^i depend on the v^i at most linearly, and on no derivatives of the v^i.*

Since we are concerned with the one-dimensional case only, "divergence" here is equivalent to a total τ-derivative. In particular, then, if

$$\delta I\equiv\int L'\,d\tau'-\int L\,d\tau=\int\frac{dC}{d\tau}\,d\tau, \tag{4}$$

where L' is the *same* function of the transformed variables (3) as L is of the untransformed ones, we have

$$\frac{d}{d\tau}\left[\left(\frac{\partial L}{\partial v^i}v^i - L\right)\delta\tau - \frac{\partial L}{\partial v^i}\delta x^i + C\right] = L_i(\delta x^i - v^i\,\delta\tau), \tag{5}$$

where summation over i is understood.

It should be noted that the statement of the theorem concerns the expressions (5), which are a direct consequence of the invariance (up to a divergence) of (1) under (3), and *not* any conservation laws; such laws only follow from (5) if the Euler-Langrange equations

$$L_i = 0 \tag{6}$$

are satisfied. However, NOETHER'S theorem does *not* claim to make a statement on the total number of conservation laws following from invariance (up to a divergence) under G_p, but (as explicitly noted by NOETHER herself) only on the number of linear combinations of the L_i which become divergences, and indeed it is possible in certain cases to obtain more than p conservation laws [20]. Furthermore, we can extend NOETHER'S theorem both by considering transformations involving functionals [19], and by allowing conserved quantities which contain functionals; a simple example of the latter case is discussed below.

We now consider the case of Newtonian point mechanics in the Lagrangean form [21]. We restrict ourselves to Cartesian coordinates and denote the position vector of the j-th particle by $\boldsymbol{r}_j$, with components x_j^i, and we now use x_j^i ($i=1, 2, 3$, $j=1 \ldots n$) in place of the N quantities x^i. Furthermore, we identify τ with the time t. Then

$$\begin{aligned} &L = T - V, \\ &T = \tfrac{1}{2}\sum_{j=1}^{n} m_j \boldsymbol{v}_j^2, \qquad \boldsymbol{v}_j \equiv \frac{d\boldsymbol{r}_{ij}}{dt}, \\ &V = \tfrac{1}{2}\sum_{i \neq j}\sum g_i\, g_j\, V_{ij}(r_{ij}), \qquad r_{ij} \equiv |\boldsymbol{r}_i - \boldsymbol{r}_j|, \end{aligned} \tag{7}$$

where the g_i are constants characteristic of the interaction, which are introduced explicitly for later convenience.

The infinitesimal transformations of the Galilei group are of the form (3), and the customary choice of ten independent transformations is

$$\delta t = \sigma, \qquad \delta x_j^i = 0, \tag{8}$$

$$\delta t = 0, \qquad \delta x_j^i = \alpha^i, \tag{9}$$

$$\delta t = 0, \qquad \delta x_j^i = \beta_l^i\, x_j^l, \qquad \beta_l^i = -\beta_i^l, \tag{10}$$

$$\delta t = 0, \qquad \delta x_j^i = \gamma^i\, t, \tag{11}$$

where σ, α^i, β^i, γ^i are ten independent infinitesimal parameters. The transformations (8)–(10) clearly leave L invariant, whereas for (11) we obtain

$$\delta J = \int \frac{dC}{dt}\, dt, \qquad C = \sum_j m_j \gamma^i x_j^i. \tag{12}$$

Using Eq. (5) together with (6), we get the conservation laws corresponding to (8)–(11)

$$\frac{dE}{dt} = 0, \qquad E \equiv \sum_j \boldsymbol{p}_j \cdot \boldsymbol{v}_j - L, \qquad p_j^i \equiv \frac{\partial L}{\partial v_j^i}, \tag{13}$$

$$\frac{d\boldsymbol{P}}{dt} = 0, \qquad \boldsymbol{P} \equiv \sum_j \boldsymbol{p}_j, \tag{14}$$

$$\frac{d\boldsymbol{J}}{dt} = 0, \qquad \boldsymbol{J} \equiv \sum_j \boldsymbol{r}_j \times \boldsymbol{p}_j, \tag{15}$$

$$\frac{d\boldsymbol{M}}{dt} = 0, \qquad \boldsymbol{M} \equiv \sum_j m_j \boldsymbol{r}_j - \boldsymbol{P} t. \tag{16}$$

The first three relations of course express the conservation of energy and of linear and angular momentum. Eq. (16) can be rewritten, introducing the (constant) total mass M and the center-of-mass coordinate $\boldsymbol{R}$,

$$\frac{d\boldsymbol{M}}{dt} = 0, \quad \boldsymbol{M} \equiv M\boldsymbol{R} - \boldsymbol{P} t, \quad M\boldsymbol{R} \equiv \sum_j m_j \boldsymbol{r}_j, \quad M \equiv \sum_j m_j, \tag{17}$$

or, using (14),

$$\frac{d\boldsymbol{R}}{dt} \equiv \frac{\boldsymbol{P}}{M} = \text{const.}, \tag{18}$$

which is the familiar form of the center-of-mass theorem.

If the potential V_{ij} is proportional to r_{ij}^{-1}, the infinitesimal transformation ("scale transformation")

$$\delta t = \alpha t, \qquad \delta x_j^i = \alpha x_j^i \tag{19}$$

leads to

$$\delta I = \alpha \int T\, dt. \tag{20}$$

Although the integrand is not a total derivative of a function, it can be written as a total derivative of a functional

$$C(t) = \int^t T(t')\, dt'. \tag{21}$$

Then Eqs. (5) and (6) lead to

$$\frac{dD}{dt} = 0, \qquad D \equiv E t - \sum_j \boldsymbol{p}_j \cdot \boldsymbol{r}_j + \int^t T(t')\, dt', \tag{22}$$

which is equivalent to

$$E = \frac{d}{dt} \sum_j \boldsymbol{p}_j \cdot \boldsymbol{r}_j - T, \tag{23}$$

an equation usually used as the starting point in deriving the virial theorem. Similar equations can be obtained for other potentials [22]; for V_{ij} proportional to r_{ij}^{-2}, one obtains Eq. (22) without the last term [23].

While "conserved quantities" such as (22) involving functionals appear to be somewhat artificial in Newtonian mechanics, they arise quite naturally in special relativistic particle dynamics (as noted before), as well as in a generalized Galilei-invariant dynamics, which are considered in the following sections.

3 *Galilei- and Lorentz-Invariant Variational Principles*

In the preceding section, the conservation laws were stated in three-dimensional notation. For the following, it is more convenient to use a four-dimensional notation for the Galilei- as well as the Lorentz-invariant case [9]. We consider the group of transformations of the Cartesian space coordinates x^1, x^2, x^3 and the time $t = x^0$

$$x'^{\mu} = \alpha^{\mu}_{\rho}\, x^{\rho} + \xi^{\mu}. \tag{24}$$

Here and in the following, summation over repeated indices is understood, and Greek indices always range from 0 to 3.

The full inhomogeneous Lorentz group (also called Poincaré group) is the group of transformations (24), restricted by the condition [24]

$$\eta_{\mu\nu}\, \alpha^{\mu}_{\rho}\, \alpha^{\nu}_{\sigma} = \eta_{\rho\sigma}, \tag{25 L}$$

where $\eta_{\mu\nu}$ is a nonsingular symmetric tensor with signature -2. We can define its inverse by

$$\eta_{\mu\rho}\, \eta^{\rho\nu} = \delta^{\nu}_{\mu} \tag{26 L}$$

and shall take the nonvanishing components of these tensors to be

$$\begin{aligned} \eta_{00} &= 1, \quad \eta_{11} = \eta_{22} = \eta_{33} = -c^{-2}, \\ \eta^{00} &= 1, \quad \eta^{11} = \eta^{22} = \eta^{33} = -c^{2}. \end{aligned} \tag{27 L}$$

The full inhomogeneous Galilei group is the group of transformations (24), restricted by the conditions

$$\begin{aligned} g_{\mu\nu}\, \alpha^{\mu}_{\rho}\, \alpha^{\nu}_{\sigma} &= g_{\rho\sigma}, \\ h^{\mu\nu}\, \alpha^{\rho}_{\mu}\, \alpha^{\sigma}_{\nu} &= h^{\rho\sigma}, \end{aligned} \tag{25 G}$$

where we can choose as the nonvanishing components of $g_{\mu\nu}$ and $h^{\mu\nu}$

$$g_{00} = 1, \quad h^{11} = h^{22} = h^{33} = -1. \tag{27 G}$$

Thus these tensors are singular and have no inverses; they satisfy the equation

$$g_{\mu\rho}\, h^{\rho\nu} = 0. \tag{26G}$$

Because of this the two sets of relations (25G) had to be postulated independently, whereas (25L) implies a similar condition for $\eta^{\mu\nu}$.

Eqs. (25) and (27) imply that the Jacobian J of the transformation equals ± 1 in both cases. Thus both the Lorentz and the Galilei group consist of four parts, corresponding to the four combinations of the signs of J and of α_0^0. The part with $J = \operatorname{sign} \alpha_0^0 = 1$ forms a subgroup, the proper orthochronous Lorentz and Galilei group, respectively. The physical requirement of invariance under this subgroup is more basic than that under the full group, because it is this subgroup which corresponds to uniform relative motion of frames of reference without reflections and time reversal, i.e. which expresses the equivalence of all inertial systems.

If we restrict ourselves to orthochronous Galilei transformations, the quantity

$$w_\mu = (1, 0, 0, 0) \tag{28G}$$

transforms like a covariant vector, with numerically the same components in all coordinate systems; under antichronous transformations (for which $\alpha_0^0 = -1$) it changes sign. We have

$$h^{\mu\rho}\, w_\rho = 0, \qquad g_{\mu\nu} = w_\mu\, w_\nu. \tag{29G}$$

The space of the Lorentz group is metric, with the metric tensor $\eta_{\mu\nu}$ and the four-dimensional distance

$$\Delta s \equiv (\eta_{\mu\nu}\, \Delta x^\mu\, \Delta x^\nu)^{\frac{1}{2}}. \tag{30L}$$

For the Galilei group, we could also introduce such a distance, using $g_{\mu\nu}$ as a metric tensor, by

$$\Delta s \equiv (g_{\mu\nu}\, \Delta x^\mu\, \Delta x^\nu)^{\frac{1}{2}}. \tag{30G}$$

However, this metric is singular and thus the space is not Riemannian. Therefore, unlike $\eta_{\mu\nu}$ and its inverse $\eta^{\mu\nu}$, $g_{\mu\nu}$ and $h^{\mu\nu}$ can not be used to lower and raise indices reversibly, and in general co- and contravariant vectors are distinct quantities. The separation (30G) actually refers to a pure time interval because of the form (27G) of $g_{\mu\nu}$, and assigns a separation zero to any two simultaneous events.

We now consider the world line of a particle j whose four coordinates z_j^ρ can be expressed as functions of an arbitrary parameter. It is convenient to take as this parameter the "proper time" τ_j defined by

$$d\tau_j \equiv (\eta_{\mu\nu}\, dz_j^\mu\, dz_j^\nu)^{\frac{1}{2}} \tag{31L}$$

$$d\tau_j \equiv (g_{\mu\nu}\, dz_j^\mu\, dz_j^\nu)^{\frac{1}{2}}, \tag{31G}$$

where the Galilean τ_j differs from $z_j^0 = \pm t$ at most by an additive constant.

Using (11), we can define a contravariant four-velocity v_j^ρ and four-acceleration a_j^ρ by

$$v_j^\rho \equiv \frac{dz_j^\rho}{d\tau_j}, \qquad a_j^\rho \equiv \frac{d^2 z_j^\rho}{d\tau_j^2}. \tag{32}$$

In the Lorentz case the correspondent covariant quantities are given by $\eta_{\rho\sigma} v_j^\sigma$ and $\eta_{\rho\sigma} a_j^\sigma$. In the Galilei case, we can not automatically define the corresponding quantities. However, the vector w_μ defined in Eq. (28 G) equals $g_{\mu\rho} v_j^\rho$, and thus can be taken formally as the covariant vector corresponding to v_j^ρ, although it is not possible to regain v_j^ρ from it. It will be convenient for conciseness in our later calculations to introduce the notation

$$u_{j\rho} \equiv v_{j\rho} = \eta_{\rho\sigma} v_j^\sigma, \tag{33 L}$$

$$u_{j\rho} \equiv w_\rho; \tag{33 G}$$

then we have in both cases

$$u_{j\rho} v_j^\rho = 1, \qquad u_{j\rho} a_j^\rho = 0, \tag{34}$$

which justifies our interpretation of w_ρ. Furthermore, it was shown in Ref. [9] that we can define a quantity

$$a_{j\rho} \equiv (\boldsymbol{a}_j \cdot \boldsymbol{v}_j, -\boldsymbol{a}_j), \tag{35 G}$$

which transforms as a covariant vector and satisfies

$$a_{j\rho} v_j^\rho = 0 \tag{36}$$

just like the corresponding quantity in the Lorentz case, and from which it *is* possible to regain a_j^ρ.

Now we consider *two* events z_i^ρ and z_j^ρ, with

$$s_{ij}^\rho(\tau_i, \tau_j) \equiv z_i^\rho(\tau_i) - z_j^\rho(\tau_j), \tag{37}$$

and the corresponding four-velocities v_i^ρ and v_j^ρ. In the Lorentz case, we can define the following independent invariants formed from these four quantities:

$$\begin{aligned} s_{ij} &\equiv (\eta_{\mu\nu} s_{ij}^\mu s_{ij}^\nu)^{\frac{1}{2}}, & \omega_{ij} &\equiv v_{i\rho} v_j^\rho, \\ \kappa_i &\equiv s_{ji}^\rho v_{i\rho}, & \kappa_j &\equiv s_{ij}^\rho v_{j\rho}. \end{aligned} \tag{38 L}$$

In the Galilei case we also have four independent invariants [9, 18]

$$\begin{aligned} t_{ij} &\equiv w_\rho s_{ij}^\rho, \qquad v_{ij} \equiv [-\tfrac{1}{2}(k_{i\rho\sigma} + k_{j\rho\sigma})(v_i^\rho - v_j^\rho)(v_i^\sigma - v_j^\sigma)]^{\frac{1}{2}}, \\ S_{ij} &\equiv [-\tfrac{1}{2}(k_{i\rho\sigma} + k_{j\rho\sigma}) S_{ij}^\rho S_{ij}^\sigma]^{\frac{1}{2}}, \\ K_{ij} &\equiv -\tfrac{1}{2}(k_{i\rho\sigma} + k_{j\rho\sigma}) S_{ij}^\rho (v_i^\sigma - v_j^\sigma), \end{aligned} \tag{38 G}$$

where

$$k_{i\rho\sigma} \equiv \varepsilon_{\rho\alpha\beta\gamma}\, \varepsilon_{\sigma\mu\nu\lambda}\, v_i^\alpha\, v_i^\mu\, h^{\beta\nu}\, h^{\gamma\lambda}, \qquad S_{ij}^\rho \equiv s_{ij}^\rho - \tfrac{1}{2}(v_i^\rho + v_j^\rho)\, t_{ij},$$

and $\varepsilon_{\rho\alpha\beta\gamma}$ is the Levi-Civita tensor density; these are more conveniently written in three-dimensional notation as

$$\begin{aligned} &t_{ij} \equiv t_i - t_j, && v_{ij} \equiv |\boldsymbol{v}_i - \boldsymbol{v}_j|, \\ &S_{ij} \equiv |\boldsymbol{r}_i - \boldsymbol{r}_j - \tfrac{1}{2}(\boldsymbol{v}_i + \boldsymbol{v}_j)\, t_{ij}|, && K_{ij} \equiv [\boldsymbol{r}_i - \boldsymbol{r}_j - \tfrac{1}{2}(\boldsymbol{v}_i + \boldsymbol{v}_j)\, t_{ij}] \cdot (\boldsymbol{v}_j - \boldsymbol{v}_i). \end{aligned} \tag{38'G}$$

If the events are simultaneous, S_{ij} reduces to the usual three-dimensional separation r_{ij}.

We could construct other invariants using higher derivatives or more than two events. However, in the following we shall restrict ourselves to the mechanics of particles involving only two-body interactions depending at most on velocities in the Lagrangean (and thus to at most acceleration-dependent forces). We do *not* assume that the interactions are symmetric in the particles.

We wish to investigate equations of motion derivable from a variational principle which has the standard form of Newtonian or of relativistic mechanics. In four-dimensional notation they can be written

$$m_j\, a_j^\rho = \frac{d P_j^\rho}{d\tau_j} = F_j^\rho, \qquad P_j^\rho \equiv m_j\, v_j^\rho, \qquad j = 1 \ldots n. \tag{39}$$

The restriction to two-body interactions implies

$$F_j^\rho = \sum_{k \neq j} F_{jk}^\rho. \tag{40}$$

Because of Eqs. (34) we have

$$F_j^\rho\, u_{j\rho} = 0, \tag{41}$$

where in three-dimensional notation

$$F_j^\rho \equiv \left[\frac{\boldsymbol{F}_j \cdot \boldsymbol{v}_j}{c^2} \left(1 - \frac{\boldsymbol{v}_j^2}{c^2}\right)^{-\frac{1}{2}}, \boldsymbol{F}_j \left(1 - \frac{\boldsymbol{v}_j^2}{c^2}\right)^{-\frac{1}{2}} \right], \tag{42L}$$

$$F_j^\rho \equiv [0, \boldsymbol{F}_j]. \tag{42G}$$

Eqs. (39) are equivalent to (see Ref. [9])

$$m_j\, a_{j\rho} = \frac{d P_{j\rho}}{d\tau_j} = F_{j\rho}, \qquad F_{j\rho}\, v_j^\rho = 0, \tag{43}$$

where [25]

$$F_{j\rho} \equiv \eta_{\rho\sigma}\, F_j^\sigma, \qquad P_{j\rho} \equiv \eta_{\rho\sigma}\, P_j^\sigma, \tag{44L}$$

$$F_{j\rho} \equiv (\boldsymbol{F}_j \cdot \boldsymbol{v}_j, -\boldsymbol{F}_j), \qquad P_{j\rho} \equiv (\tfrac{1}{2} m_j\, \boldsymbol{v}_j^2, -\boldsymbol{p}_j). \tag{44G}$$

To obtain equations of the form (39) or (43) with (40) we must start in the Lorentz case from a variational principle of the form

$$I_1 \equiv -\sum_{i<j}\sum g_i\, g_j \iint_{-\infty}^{\infty} U_{ij}(s_{ij}, \omega_{ij}, \kappa_i, \kappa_j)\, d\tau_i\, d\tau_j, \tag{45L}$$

where the g_i are constants, and the variation has to be performed subject to the subsidiary condition (34) with (33L). This is achieved as usual by adding

$$I_2 \equiv -\sum_j \int_{-\infty}^{\infty} M_j(\tau_j)(v_{j\rho}\, v_j^\rho)^{\frac{1}{2}}\, d\tau_j \tag{46L}$$

to (45L), where the M_j are Lagrangean multipliers which have to be determined from the equations resulting from the variation of $I_1 + I_2$. The square root in (46L) was introduced for later convenience, and is necessary if a parameter-invariant integral is desired.

In the Galilei case we must start from

$$I_1 + I_3 \equiv -\sum_{i<j}\sum g_i\, g_j \iint_{-\infty}^{\infty} U_{ij}(t_{ij}, v_{ij}, S_{ij}, K_{ij})\, d\tau_i\, d\tau_j + \sum_j \int_{-\infty}^{\infty} \tfrac{1}{2} m_j\, \mathbf{v}_j^2\, d\tau_j \tag{45G}$$

subject to the subsidiary condition (34) with (33G). Therefore we have to add

$$I_2 \equiv -\sum_j \int_{-\infty}^{\infty} M_j(\tau_j)\, w_{j\rho}\, v_j^\rho\, d\tau_j \tag{46G}$$

to (45G). We note, using (33), that

$$\delta I_2 = -\sum_j \int_{-\infty}^{\infty} M_j\, u_{j\rho}\, \delta v_j^\rho\, d\tau_j \tag{47}$$

in both cases. The signs in (45) and (46) are chosen for convenience and to maintain the analogy to the Newtonian principle (7).

For conciseness in later use it will be convenient to define "adjunct potentials" V_j by

$$V_j(z_j^\rho, v_j^\rho) \equiv \sum_{k<j} g_k \int_{-\infty}^{\infty} U_{kj}\, d\tau_k + \sum_{k>j} g_k \int_{-\infty}^{\infty} U_{jk}\, d\tau_k \tag{48}$$

in terms of the "kernels" U_{ij} of the variational principles, which are not assumed to be necessarily symmetric in i and j.

Now we carry out the variation of (45), adding (46); Eqs. (6) with (2) then give

$$-g_j \frac{\partial V_j}{\partial z_j^\rho} + \frac{d}{d\tau_j}\left[g_j \frac{\partial V_j}{\partial v_j^\rho} + M_j\, u_{j\rho}\right] = 0, \tag{49L}$$

$$-g_j \frac{\partial V_j}{\partial z_j^\rho} + \frac{d}{d\tau_j}\left[g_j \frac{\partial V_j}{\partial v_j^\rho} + M_j\, u_{j\rho}\right] - m_j\, a_j^\rho = 0, \tag{49G}$$

where the mixture of co- and contravariant indices in (49 G) is due to the fact that I_3 is not fully Galilei-invariant. On contraction of these equations with v_j^ϱ we obtain expressions for $dM_j/d\tau_j$ which can be integrated to yield

$$M_j = g_j \left(V_j - v_j^\sigma \frac{\partial V_j}{\partial v_j^\sigma} \right) + m_j, \tag{50 L}$$

$$M_j = g_j \left(V_j - v_j^\sigma \frac{\partial V_j}{\partial v_j^\sigma} \right) + \tfrac{1}{2} m_j \boldsymbol{v}_j^2 + \text{const.}, \tag{50 G}$$

where the form of the constant of integration in (50 L) is chosen for convenience, and is irrelevant in (50 G) due to the form of (49 G).

Thus, using the notation (33), the equations of motion (49) can be written

$$\frac{dP_{j\rho}}{d\tau_j} = g_j \frac{\partial V_j}{\partial z_j^\varrho} - g_j \frac{d}{d\tau_j} \left[\frac{\partial V_j}{\partial v_j^\varrho} + u_{j\rho} \left(V_j - v_j^\sigma \frac{\partial V_j}{\partial v_j^\sigma} \right) \right]. \tag{51}$$

Apart from the additional complication introduced by the last term in (45 G), the developments of the Galilei- and the Lorentz-invariant case were completely parallel. The reason for this complication is that due to the properties of w_μ, the contribution (46 G) of the constraint (34 G) is not sufficient to provide a nontrivial spatial part in the equations of motion; on the other hand, as shown in Section 2, I_3 is not fully invariant under the Galilei group, but [for the transformation (11)] only up to a divergence. I_1 is invariant under the respective proper orthochronous groups by the properties of the invariants in both cases.

The equations of motion (51) relate the acceleration of the j-th particle at a given time to the values of the variables of the *other* particles at *different* times (except in the Newtonian case $t_{ij}=0$). In the Lorentz case, the separation of the particles may be time- or space-like, depending on the properties of U_{ij}; no conceptional difficulties are introduced by this, as long as we are dealing with a *closed* system of particles [7], a restriction inherent also in the Newtonian mechanics of particle systems.

The generalization to noninstantaneous Galilean interactions was introduced in Ref. [18]. The Lorentz case is a generalization of the Fokker-type variational principles designed for providing a theory of direct particle interactions corresponding to a field theory [10–12], which involve only time-like four-dimensional separations. Space-like separations by means of relations involving all invariants (38 L) were first considered by HAVAS and PLEBAŃSKI [26], and by means of s_{ij} alone later by VAN DAM and WIGNER [27].

The special case of theories possessing an "adjunct field theory" can be obtained from the principles (45) provided the "adjunct potential"

V_j given by (48) can be separated into a sum of terms which are products of two factors, one of which involves only the coordinates z_j^ρ of particle j.

In the Galilei case, the form (38′G) of the invariants implies that this is possible *only* if $t_{ij}=0$ and the U_{ij} are functions of S_{ij} alone. Then we obtain the usual Newtonian theory, and the variational principle (45 G) with (46 G) is equivalent to (1) with (7). V_j can then always be considered as the solution of a partial differential equation, the field equation of the adjunct field theory.

In the Lorentz case, the form (38 L) of the invariants implies that U_{ij} consists of a sum of terms each of which contains ω_{ij} only in the form ω_{ij}^l, where l is an integer. Concentrating for simplicity only on one of these with a particular l, we have then

$$V_j^{(l)}(z_j^\rho, v_j^\rho) = v_{j\alpha}\, v_{j\beta} \ldots v_{j\lambda}\, \varphi_j^{\alpha\beta\ldots\lambda}(z_j^\rho),$$

$$\varphi_j^{\alpha\beta\ldots\lambda} = \sum_{k<j} g_k \int_{-\infty}^{\infty} v_k^\alpha\, v_k^\beta \ldots v_k^\lambda\, \varphi_{kj}(s_{jk}, \kappa_k)\, d\tau_k \qquad (52\,\mathrm{L})$$

$$+ \sum_{k>j} g_k \int_{-\infty}^{\infty} v_k^\alpha\, v_k^\beta \ldots v_k^\lambda\, \varphi_{jk}(s_{jk}, \kappa_k)\, d\tau_k .$$

We could simply consider as the field equation of the adjunct field theory the partial differential equation which has $\varphi_j^{\alpha\beta\ldots\lambda}(x^\rho)$ as a solution. However, customarily one only considers field equations which consist of a homogeneous part involving $\varphi_j^{\alpha\beta\ldots\lambda}(x^\rho)$ alone, and a source part depending on the z_k^ρ and v_k^ρ. A sufficient, but not necessary, condition for this is that the φ_{jk} do not depend on the κ_k and, considered as functions G_k of $x^\rho - z_k^\rho$ rather than $z_j^\rho - z_k^\rho$, are Green functions of some linear partial differential equation, i.e. that

$$\mathscr{L} G_k[\eta_{\mu\nu}(x^\mu - z_k^\mu)(x^\nu - z_k^\nu)] = 4\pi\, \delta^4(x^\rho - z_k^\rho), \qquad (53\,\mathrm{L})$$

where $\mathscr{L}$ is a linear differential operator and δ^4 is a four-fold product of Dirac δ-functions. Then

$$\mathscr{L} \varphi_j^{\alpha\beta\ldots\lambda}(x^\rho) = 4\pi \sum_{k\neq j} \int_{-\infty}^{\infty} v_k^\alpha\, v_k^\beta \ldots v_k^\lambda\, \delta^4[x^\rho - z_k^\rho(\tau_k)]\, d\tau_k \equiv 4\pi\, j_j^{\alpha\beta\ldots\lambda}, \qquad (54\,\mathrm{L})$$

where $j_j^{\alpha\beta\ldots\lambda}$ is the source density of the adjunct field of the j-th particle.

Special cases of theories of direct particle interactions with adjunct field theories are FOKKER's principle of electrodynamics [10] ($l=1$, $G_k=\delta$), and the principles of scalar or vector mesodynamics [11] ($l=0$ or 1).

However, we are by no means restricted to such theories. In particular, there exist many distinct theories with space- as well as time-like interactions which are relativistic generalizations of the same Newtonian theory [28, 29].

4 *Conservation Laws*

The equations of motion (51) are the consequence of a variational principle (45) with (46), where the interaction term is invariant under the 10-parameter group (24) with (25), and I_3 in (45G) is invariant up to a divergence. Therefore NOETHER's theorem assures us of the existence of 10 conservation laws.

The interaction term I_1 in (47) has essentially the same form as the one used by DETTMAN and SCHILD in deriving conservation laws [30]. These authors aimed at a theory of variable mass, and therefore included terms $j=i$ in their summation; furthermore, they only considered symmetric Lorentz-invariant U_{ij}'s and no subsidiary conditions (46). The modifications necessary to take care of the Galilei-invariant case as well have been discussed elsewhere [18] and those taking account of the other differences mentioned are slight. Caution is required only because in the Galilei case we can not raise and lower indices at will, due to the absence of a metric. Furthermore, unlike the Lorentz group, the Galilei group does not allow a treatment of the spatial rotations and the "boosts" (leading to conservation of angular momentum and the center-of-mass theorem) together as four-dimensional rotations. Nevertheless, formally a unified treatment is possible, and for brevity only this treatment is presented here; the separate treatment, which shows a closer analogy to that of the Newtonian case of Section 2, is discussed in Ref. [18].

We consider the variational principles (45) with (46), where, however we now consider limits τ_j^* and τ_j^{**} which are arbitrary except that τ_j^{**} always refers to a later time than τ_j^*, and that for the variations leading to the equations of motion (considered in the previous section) we take the limits $\tau_j^* \to -\infty$, $\tau_j^{**} \to +\infty$, for all j.

To obtain the conservation laws we use the infinitesimal transformations associated with the groups (24) with (25). For the law of conservation of energy-momentum we consider the (constant) infinitesimal space-time translations

$$\delta z_j^\mu = \varepsilon^\mu, \tag{55}$$

for which the total variation of (45) with (46) vanishes, since the integrands of these expressions are invariant under (55). Then δI_1 can be treated just as in Ref. [30], apart from taking due account of the possible lack of symmetry in i and j. The evaluation of δI_2 and of δI_3 is straightforward.

Inserting the equation of motion (51) into the resulting expression and combining it with δI_1, we then obtain

$$\frac{\partial P_\mu}{\partial \tau_j}=0, \quad P_\mu(\tau_1\dots\tau_n)\equiv\sum_j\left\{P_{j\mu}(\tau_j)+g_j\left[\frac{\partial V_j}{\partial v_j^\mu}+u_{j\mu}\left(V_j-v_j^\sigma\frac{\partial V_j}{\partial v_j^\sigma}\right)\right]\right\}$$
$$+\sum_{i<j}\sum g_i\,g_j\left(\int_{\tau_i}^{\infty}\int_{-\infty}^{\tau_j}-\int_{-\infty}^{\tau_i}\int_{\tau_j}^{\infty}\right)\frac{\partial U_{ij}}{\partial s_{ij}^\mu}\,d\tau_i\,d\tau_j \tag{56}$$

independent of where on the world lines these quantities are evaluated [31].

To obtain the laws of conservation of angular momentum and the center-of-mass theorem we consider the infinitesimal transformations

$$\delta z_j^\rho=H^{\rho\mu}\,\varepsilon_{\mu\nu}\,z_j^\nu, \qquad \varepsilon_{\mu\nu}=-\varepsilon_{\nu\mu}, \tag{57}$$

where the $\varepsilon_{\mu\nu}$ are arbitrary constant parameters, and where we have introduced the notation

$$H^{\rho\mu}\equiv\eta^{\rho\mu}, \tag{58 L}$$

$$H^{\rho\mu}\equiv h^{\rho\mu}, \tag{58 G}$$

to allow a common treatment of the Galilei and the Lorentz case. The integrands of I_1 and I_2 are again invariant under the variations (57) and can be treated in analogy to those for the transformation (55). However, I_3 is only invariant up to a divergence; explicitly we find

$$\delta I_3=\sum_j\int_{\tau_j^*}^{\tau_j^{**}}\frac{d}{d\tau_j}(m_j\,\varepsilon_{m0}\,z_j^m)\,d\tau_j. \tag{59 G}$$

We finally obtain in both cases

$$\frac{\partial L^{\mu\nu}}{\partial \tau_j}=0,$$
$$L^{\mu\nu}(\tau_1\dots\tau_n)\equiv\sum_j\left\{\left\{P_j^\mu+g_j\left[\frac{\partial V_j}{\partial v_j^\rho}+u_{j\rho}\left(V_j-v_j^\sigma\frac{\partial V_j}{\partial v_j^\sigma}\right)\right]H^{\rho\mu}\right\}z_j^\nu\right.$$
$$\left.-\left\{P_j^\nu+g_j\left[\frac{\partial V_j}{\partial v_j^\rho}+u_{j\rho}\left(V_j-v_j^\sigma\frac{\partial V_j}{\partial v_j^\sigma}\right)\right]H^{\rho\nu}\right\}z_j^\mu\right\}$$
$$+\sum_{i<j}\sum\left(\int_{\tau_i}^{\infty}\int_{-\infty}^{\tau_j}-\int_{-\infty}^{\tau_i}\int_{\tau_j}^{\infty}\right)d\tau_i\,d\tau_j \tag{60}$$
$$\cdot\left[\frac{\partial U_{ij}}{\partial s_{ij}^\rho}(H^{\rho\mu}\,s_{ij}^\nu-H^{\rho\nu}\,s_{ij}^\mu)+\frac{\partial U_{ij}}{\partial v_i^\rho}(H^{\rho\mu}\,v_i^\nu-H^{\rho\nu}\,v_i^\mu)\right.$$
$$\left.+\frac{\partial U_{ij}}{\partial v_j^\rho}(H^{\rho\mu}\,v_j^\nu-H^{\rho\nu}\,v_j^\mu)\right],$$

where the spatial components L^{mn} are the components of the angular momentum $\boldsymbol{J}$ analogous to (15) and the components L^{m0} are those of the vector $\boldsymbol{M}$ of the center-of-mass theorem analogous to (16).

For certain forms of U_{ij} we can obtain additional conservation laws. Here we consider only the infinitesimal transformation

$$\delta z_j^\mu = \alpha z_{ij}^\mu \tag{61}$$

[equivalent to (19)], implying $\delta\tau_j = \alpha\tau_j$, which leaves invariant any I_1 given by (45) with symmetric interactions of the form

$$I_1 = \sum_{i<j}\sum g_i g_j \iint \delta(s_{ij}^2) f_{ij}(\omega_{ij})\, d\tau_i\, d\tau_j, \tag{62L}$$

$$I_1 = \sum_{i<j}\sum g_i g_j \iint \delta(t_{ij}) \frac{f_{ij}(v_{ij})}{S_{ij}}\, d\tau_i\, d\tau_j, \tag{62G}$$

where the f_{ij} are arbitrary functions. Proceeding as before we obtain

$$\frac{\partial D}{\partial \tau_j} = 0,$$

$$\begin{aligned} D(\tau_1 \ldots \tau_n) \equiv \sum_j \left\{ P_{j\mu} + g_j \left[\frac{\partial V_j}{\partial v_j^\mu} + u_{j\mu}\left(V_j - v_j^\sigma \frac{\partial V_j}{\partial v_j^\sigma} \right) \right] \right\} z_j^\mu \\ - \sum_j m_j \tau_j - \tfrac{1}{2} \sum_{i<j}\sum g_i g_j \left(\int_{\tau_i}^{\infty} \int_{-\infty}^{\tau_j} - \int_{-\infty}^{\tau_i} \int_{\tau_j}^{\infty} \right) \frac{\partial \delta(s_{ij}^2)}{\partial s_{ij}^\mu} \\ \cdot f_{ij} (z_i^\mu + z_j^\mu)\, d\tau_i\, d\tau_j, \end{aligned} \tag{63L}$$

$$\frac{\partial D}{\partial \tau_j} = 0,$$

$$\begin{aligned} D(\tau_1 \ldots \tau_n) \equiv \sum_j \left\{ P_{j\mu} + g_j \left[\frac{\partial V_j}{\partial v_j^\mu} + u_{j\mu}\left(V_j - v_j^\sigma \frac{\partial V_j}{\partial v_j^\sigma} \right) \right] \right\} z_j^\mu \\ + \sum_j \int_{-\infty}^{\tau_j} \tfrac{1}{2} m_j \boldsymbol{v}_j^2\, d\tau_j - \tfrac{1}{2} \sum_{i<j}\sum g_i g_j \left(\int_{\tau_i}^{\infty} \int_{-\infty}^{\tau_j} - \int_{-\infty}^{\tau_i} \int_{\tau_j}^{\infty} \right) \\ \cdot \frac{\partial}{\partial s_{ij}^\mu} \left[\frac{\delta(t_{ij})}{S_{ij}} \right] f_{ij} (z_i^\mu + z_j^\mu)\, d\tau_i\, d\tau_j. \end{aligned} \tag{63G}$$

For $f_{ij} = 1$, Eq. (63G) is equivalent to (22). For $f_{ij} = \omega_{ij}^l$, the interaction (62L) is equivalent to (52L) with the Green function $\delta(s_{ij}^2)$ such that the adjunct field satisfies the wave equation; for $l = 1$ (the case of FOKKER'S principle of electrodynamics [10]), Eq. (63L) was derived recently [22], and represents the action-at-a-distance analog of a conservation law found some time ago for a system of particles and field [32].

All the conservation laws considered involve functionals which depend on the history of the particles, except for the case of instantaneous interactions ($t_{ij}=0$) for the Galilean conservation laws (56) and (60), where they become functions of the particle variables at a single time. As shown in Ref. [18], it is possible to choose more general forms of Galilei-invariant interactions for which these conservation laws do not involve functionals, but only particle variables at definite times; these are different for each particle, however. This is also possible for Lorentz-invariant interactions, but apparently only for a system of two particles. The Galilei-invariant interaction are of the form

$$\begin{aligned} I_1 &\equiv -\sum_{i<j}\sum g_i\, g_j \iint_{-\infty}^{\infty} \delta(t_{ij}-T_{ij})\, g_{ij}(t_{ij}, v_{ij}, S_{ij}, K_{ij})\, d\tau_i\, d\tau_j, \\ T_{ij} &\equiv T_i - T_j, \end{aligned} \tag{64G}$$

where the T_i are constants, and the g_{ij} are arbitrary functions. The Lorentz-invariant interactions are of the form

$$\begin{gathered} I_1 \equiv -g_1\, g_2 \iint_{-\infty}^{\infty} \delta(s_{12}^2)\, \Theta(z_1^0 - z_2^0)\, g_{12}(s_{12}, \omega_{12}, \kappa_1, \kappa_2)\, d\tau_1\, d\tau_2, \\ \Theta(z)=1 \quad \text{for } z>0, \\ \Theta(z)=0 \quad \text{for } z<0, \end{gathered} \tag{64L}$$

where g_{12} is an arbitrary function; thus particle 2 has a retarded effect on particle 1, and 1 has an advanced effect on 2. This is a generalization of the case $g_{12}=\omega_{12}$ of an asymmetric electrodynamic interaction suggested by FOKKER [33], for which he also established the form of the energy-momentum conservation law.

It is often considered as an advantage of field theories (derivable from a variational principle) that they allow a formulation of *local* conservation laws. Such a formulation is also possible, however, for a class of action-at-a-distance theories. Such theories must possess an adjunct field theory, as discussed in Section 3. Furthermore, the adjunct fields must satisfy all the field equations of a field theory derivable from a variational principle. This is the case for any (Newtonian or special relativistic) scalar adjunct field, as it obeys only a single partial differential equation, which is derivable from the principle (45) itself. It is also possible for the vector adjunct field (52L) with $l=1$ and a Green function satisfying Eq. (53L) with $\mathscr{L}$ equal to the wave operator or the Klein-Gordon operator, since then the adjunct fields can be shown to satisfy not only Eq. (54L), but also the other set of field equations of electro-

and mesodynamics, respectively. This allows the construction of an energy-momentum tensor quadratic in the adjunct fields, and of "local" conservation laws just as in field theory, but free of its infinities [34]; of course, this tensor, just like the adjunct field itself, is not considered to have a physical significance, but only the particle interactions themselves. The main physical problems of such an approach are whether it can provide a satisfactory description of classical radiation phenomena, and whether it permits quantization. A way toward a solution of the first problem was suggested by WHEELER and FEYNMAN [35]; discussion of the second one is outside the scope of this article.

Many of the equations and results discussed here can be generalized to encompass interactions which involve transfer of electric charge (as well as of energy, momentum, and angular momentum) between particles, and particles with an intrinsic angular momentum [36].

Footnotes

1. KLEIN, F.: Vorlesungen über die Entwicklung der Mathematik im 19. Jahrhundert, vol. 2, ch. 2. Berlin: Springer 1927.
2. For an elementary derivation for the case of gravitational interaction see e.g. R. KURTH, Introduction to the mechanics of stellar systems, ch. III. New York: Pergamon Press 1957. A detailed discussion is given by A. WINTNER, The analytical foundations of celestial mechanics, ch. V. Princeton: Princeton University Press 1947.
3. NOETHER, E.: Nachr. Akad. Wiss. Göttingen, Math.-Phys. Kl. 235 (1918).
4. A simplified discussion with examples is given by E. L. HILL, Rev. Modern Phys. **23**, 253 (1951); for a more general discussion see A. TRAUTMAN, in: Gravitation, ed. by L. WITTEN, ch. 5, New York-London: John Wiley & Sons 1962, or in: Brandeis summer institute in theoretical physics, 1964, vol. 1, ch. 7. Englewood Cliffs, N. J.: Prentice-Hall, Inc. 1965.
5. BESSEL-HAGEN, E.: Math. Ann. **84**, 258 (1921). This paper includes a slightly more general formulation of Noether's theorems than stated in Ref. [3]; it too is credited to Noether. We shall use this formulation throughout.
6. For a discussion of the historical development of these concepts see MARY B. HESSE, Forces and fields. London: Thomas Nelson & Sons Ltd 1961.
7. For a brief discussion of the physical and conceptual problems see P. HAVAS, in: Proceedings of the 1964 international congress of logic, methodology and philosophy of science, p. 347. Amsterdam: North-Holland Publ. Co. 1965.
8. Surveys of recent work are given by D. G. CURRIE and T. F. JORDAN, in: Lectures in theoretical physics, vol. XA, p. 91, New York: Gordon and Breach 1968, and by P. HAVAS, in: Statistical mechanics of equilibrium and non-equilibrium, ed. by J. MEIXNER, p. 1. Amsterdam: North-Holland Publ. Co. 1965.
9. HAVAS, P.: Rev. Modern Phys. **36**, 938 (1964), and references given there.
10. First found for point charges by A. D. FOKKER, Z. Physik **58**, 386 (1929); compare also K. SCHWARZSCHILD, Nachr. Akad. Wiss. Göttingen, Math.-Phys. Kl. 126 (1903).
11. For the case of mesic interactions, see P. HAVAS, Phys. Rev. **87**, 309 (1952), and references given there.
12. Such a variational principle can be obtained also in a Lorentz-invariant approximation method in the general theory of relativity, as shown by P. HAVAS and J. N. GOLDBERG, Phys. Rev. **128**, 398 (1962).

13. See P. G. Bergmann in the article on Special Relativity, part Bl, in: Encyclopedia of physics, ed. by S. Flügge, vol. IV. Berlin-Göttingen-Heidelberg: Springer 1962.
14. Schütz, J. R.: Nachr. Akad. Wiss. Göttingen, Math.-Phys. Kl. 110 (1897).
15. Königsberger, L.: Die Principien der Mechanik. Leipzig: B. G. Teubner 1901.
16. Eisenbud, L., Wigner, E. P.: Proc. Nat. Acad. Sci. U.S.A. **27**, 281 (1941).
17. Houtappel, R. M. F., van Dam, H., Wigner, E. P.: Rev. Modern Phys. **37**, 595 (1965). — Arens, R.: J. Math. Phys. **1**, 1341 (1966). n-body forces seem to have been suggested first by G. T. Fechner, Physikalische and Philosophische Atomlehre. Leipzig: H. Mendelssohn 1864.
18. Havas, P. (to be published).
19. The presentation of this section substantially is that given by P. Havas and J. Stachel, Phys. Rev. **185**, 1636 (1969).
20. For an example with infinitely many conservation laws see H. Steudel, Z. Naturforsch. **21**A, 1826 (1966).
21. For the specific case of Newton's law of gravitation, the method and results of this section are due to Bessel-Hagen [5] and can also be found in Hill [4]. The generalization to arbitrary potentials of the form (7) is trivial.
22. Anderson, C. M., Baeyer, H. C. von (in press). It should be noted that these authors obtained such conservation laws by considering scale transformations of the masses and coupling constants in addition to those of the coordinates, which is not necessary, as just demonstrated.
23. This result is implicitly contained in H. A. Kastrup, Nucl. Phys. **58**, 561 (1964).
24. In the following, corresponding formulae of the Lorentz- and the Galilei case will be designated by L and G, respectively; formulae without a letter hold for both cases.
25. It should be noted that the Galilean and Lorentzian $F_{j\rho}$ (unlike the F_j^ρ) have different dimensions; the same difference appears in the U_{ij} of Eqs. (45) and the V_j of Eq. (48) below. The necessity of such a difference is discussed in Ref. [9].
26. Havas, P., Plebański, J.: Bull. Amer. Phys. Soc. **5**, 433 (1960); see also P. Havas, Ref. [8].
27. van Dam, H., Wigner, E. P.: Phys. Rev. **138**, B 1576 (1965); **142**, 838 (1966).
28. For the case of Coulomb's law this will be discussed elsewhere.
29. Approximately relativistic Lagrangeans (with correction terms up to order $1/c^2$) are discussed by P. Havas and J. Stachel (Ref. [19]), by H. Woodcock, Temple University Thesis, 1971, and in several papers in preparation by these authors.
30. Dettman, J. W., Schild, A.: Phys. Rev. **95**, 1057 (1954).
31. For Fokker's principle in electrodynamics (Ref. [10]), the conservation law (56) was derived by J. A. Wheeler and R. P. Feynman, Rev. Mod. Phys. **21**, 425 (1949).
32. Havas, P.: Phys. Rev. **87**, 898 (1952).
33. Fokker, A. D.: Physica **9**, 33 (1929).
34. In electrodynamics this was shown first by J. Frenkel, Z. Physik **32**, 518 (1925) and J. L. Synge, Trans. Roy. Soc. Can. **34**, 1 (1940); an alternate form of the energy-momentum tensor was suggested in Ref. [31]. The results were generalized to mesodynamics in Ref. [11].
35. Wheeler, J. A., Feynman, R. P.: Rev. Modern Phys. **17**, 157 (1945); see also P. Havas, Phys. Rev. **74**, 456 (1948); **86**, 974 (1952).
36. Havas, P.: Phys. Rev. **91**, 997 (1953), and a paper in preparation.

III. Of Fields

4 Foundations Problems in General Relativity [1]

PETER G. BERGMANN

Department of Physics, Syracuse University
Syracuse, N. Y.

Gravity has attracted the concern and the enthusiasm of a small but enduring number of physicists, which in the last two decades has even shown signs of rapid growth. I shall attempt to explain the fascination of this subject, its role within the sweep of physics research, and comment on some current activities.

ISAAC NEWTON made the theory of gravitation the paradigm of all physical theories. Using relatively modest information summarized in KEPLER'S laws of planetary motion, and realizing that the same force that governed the trajectories of the celestial bodies also is responsible for the free fall of apples on earth, he evolved a quantitative theory of the gravitational interaction of all objects possessing mass. This theory explained all known facts of gravitationally controlled motion (with the exception of MERCURY'S perihelion advance) through three centuries and, moreover, made it possible to predict details of planetary and satellite trajectories centuries ahead. Even in designing the trajectories of today's space vehicles, determining their launching parameters, and calculating required orbit corrections, space technologists have found NEWTON'S equations work perfectly.

Today's prevailing theory of gravitation, the general theory of relativity, did not grow out of any failure of NEWTON'S equations (though it explains the one slight discrepancy on MERCURY'S orbit, which had been discovered in mid-nineteenth century) [1, 2], but resulted from an attempt to reconcile the two contradictory sets of notions of space and time appropriate to NEWTON'S theories and to an account of electromagnetic phenomena, respectively.

Briefly, NEWTON considered time as universal, and as being held in common by all reasonable frames of reference. He took it for granted that the notion of simultaneity of two (distant) events was unrelated to the

[1] Supported in part by the National Science Foundation, Grant GP-19378.

state of motion of the observer, or of the instrumentation used to determine, for instance, which (if any) of two events preceded the other in time.

In electrodynamics, signals, or any other disturbances of an established condition of the field, spread at a finite speed, the speed of light. Information about two distant events can be correlated only if this information is physically propagated to a common site, where comparisons can then be made by man or instrument. For events that are not too widely separated from each other in time (specifically, they must be so nearly simultaneous that a flash announcing the occurrence of either will not reach the site of the other before that second event takes place), the motion of the instrument used for comparison determines which of these two events is considered to have taken place first. As a result, synchronization of clocks that are in relative motion, or merely distant from each other, depends on how the synchronization is performed, in a manner that can be predicted on the strength of the hypothesis that to any observer in whatever state of (uniform, rectilinear) motion the speed of light has the universal value $c = 3 \times 10^{10}$ cm sec^{-1}.

That the speed of light is indeed a universal constant is confirmed by a wealth of observational evidence. The resulting dependence of time and distance determinations on the state of motion of the observer is systematized in the so-called special theory of relativity. The corner stone of Newtonian mechanics is the instantaneous reach of forces from one massive body to another ("action at a distance"), and this reach requires absolute simultaneity. Thus two conflicting space-time structures (or models) arose from theories, each of which was phenomenally successful in its area of applicability.

It was to reconcile this conflict that EINSTEIN formulated the general theory of relativity. In the first instance this new theory deals with gravitation; it modifies NEWTON'S mechanical approach and replaces action at a distance with a new field, which mediates the interactions between massive bodies. Disturbances of a gravitational nature in this field theory will propagate at a finite rate, the same rate at which electromagnetic phenomena spread. The velocities of heavenly bodies in our solar system are sufficiently small so that the deviations from NEWTON'S laws resulting from the field interpretation are too minute in almost all instances to be observable with classical techniques. If this were all the innovation brought about by general relativity, it would have set to rest any worries about a self-contradictory space-time picture, but would not have been memorable as a new departure.

The novel idea of general relativity has been the incorporation of an ancient discovery into the foundations of gravitational theory; this is the fact that in a field of gravity all bodies undergo the same acceleration. As a result, it is not possible to separate by local experiment the influence of gravity from the possible intrinsic accelerations of a frame of reference; inertial frames of reference can be established only by reference to global considerations, i.e. in practice by the use of astronomical information. By hypothesizing that even global techniques might break down, EINSTEIN destroyed the hitherto preeminent role of the inertial frames of reference and of the transformations connecting inertial frames (the Lorentz-Poincaré group).

Thus, in the first instance, the topological manifold that we know as space-time is deprived of its previous rigidity (the MINKOWSKI universe) and endowed, instead, with the possibility of a rich geometric structure, which might be the carrier of physical properties. In this sense general relativity has fused geometry and physics. The actual structure adopted by EINSTEIN, principally on the grounds of a correspondence with the earlier MINKOWSKI geometry, was an indefinite, pseudo-Riemannian metric, which carries with it not only a quadratic norm for infinitesimal displacement vector fields, but also an affine connection and a curvature. The dynamical laws proposed by EINSTEIN consist of a connection between the local curvature and the sources of the gravitational field, that is to say masses, mass flux, and stress. As a physical theory general relativity has been completely successful: Wherever its predictions deviate from those of Newtonian physics, experiments have confirmed the presence of these "relativistic effects".

EINSTEIN'S genius enabled him to formulate the most conservative hypothesis on space-time geometry compatible with the non-existence of a rigid (pseudo-Euclidean) geometry that accounted for the known facts. There is no reason to believe, and neither did EINSTEIN himself believe, that his theory will endure without modifications as experimental techniques are increasingly refined. Even from a conceptual point of view, once the traditional space-time notions have been subjected to questioning, further generalizations almost offer themselves to our speculations.

Unified field theories have been proposed, by EINSTEIN and by others, in which the space-time manifold is endowed with a variety of geometric structures, so that one might combine the dynamics of the gravitational field with all other known physical fields, i.e. with the electromagnetic and, today, with the lepton and hadron fields. Some unified field theories go beyond the four-dimensionality of space-time and conjecture additional dimensions, which are then restricted by the assumption of isometries or near-isometries, so as to account for the observed four-dimensionality [3].

Even within general relativity there is the possibility of true singularities of space-time, i.e. regions in which the regular metric structure breaks down. Such a break-down implies that at the very least the partial differential equations that form the field laws of general relativity are not satisfied in the singular regions. Either all "local" solutions of the field laws whose continuation in space-time leads to singularities must be excluded as "unphysical", or else the true laws of nature are less restrictive than the field equations; they are equivalent to the field equations only under circumstances that do not obtain everywhere. That the highly non-linear partial differential equations of general relativity frequently, and inevitably, involve the appearance of singularities gives rise to questions as to the ultimate validity of the field equations, and their appropriateness as the mathematical formulation of the laws of nature.

Another aspect of the equations of general relativity is that at least some solutions of the field equations imply space-time topologies that deviate grossly from what is ordinarily called Euclidean manifolds. Not only the metric properties of space-time are made subject to dynamical laws but the very connectedness of that continuum. These complexities have been referred to, variously, as EINSTEIN-ROSEN bridges, as wormholes, and as Swiss cheese models, depending on the character of the space-time contortions assumed [4, 5].

One might go a step further. General relativity dispenses with rectilinear coordinate systems but retains the notion that the space-time continuum is composed of points, which in their turn form (topological) neighborhoods, or open sets. The continuous and differentiable mappings of the space-time manifold on itself imply and justify the continuing role of continuous and differentiable coordinate transformations. The laws of nature, according to general relativity, are to be invariant under this class of transformations, because intrinsically all points, and all neighborhoods of the space-time continuum are equivalent. But whereas the appearance of singularities and of exotic topologies challenges the supposed equivalence of all world points, their very existence as intrinsic objects is called into doubt by the existence of coordinate transformations that do not map point onto point, but rather field onto field [6].

To explain this point properly, it is necessary to make a distinction between several different types of (curvilinear) coordinate transformations. Presumably a coordinate transformation may be regarded as an instruction to map every point of a manifold uniquely, and reversibly, on another point. As long as the instruction makes no reference to the field at that point (or elsewhere), this is a normal point-to-point mapping, and these mappings form a group (or a groupoid). But it is also possible to conceive of an instruction that relates the point-to-point mapping to the field (metric or otherwise) defined on the manifold. As a result of that mapping, the field changes as well, according to its own transformation law (e.g. that of a tensor field); this change must be taken into consideration if two such mappings are to be performed in succession. To study the structure of the resulting group (or groupoid) one is better off thinking of the mapping of the function space of possible fields on itself, in other words the instruction maps field onto field. This more sophisticated kind of coordinate transformation contains the straight point-to-point mapping as a subgroup. The remarkable fact is that generalrelativistic theories are invariant with respect to coordinate transformations both of the straight point-to-point variety and of the more general kind.

The situation may be likened to the well-known difference, in classical mechanics, between the Lagrangian and the Hamiltonian formalism. Canonical transformations map a point in configuration space on another in configuration space, but uniquely only if we also know the values of the momentum components. That is to say, points possessing the same q-coordinates but different p-coordinates are mapped on points having different q-coordinates as well. Hence, under canonical transformations the points of configuration space lose their intrinsic identities, they do not form invariant subspaces in phase space. This is precisely the case with world points under the more general coordinate transformations: A world point (identified by its coordinate values or, perhaps, by its geometric properties) gets mapped on a definite world point only if the field is fixed; clothed with different fields it will get mapped on different world points. In this sense one may speak quite appropriately of the evanescent nature of the world point under the more general type of coordinate transformation.

One should not exaggerate the degree of disappearance of the identifiable world point. The function space mappings with respect to which the theory is invariant are still those that may be described as coordinate transformations, or as point mappings. One may, however, contemplate the possibility of theories beyond the framework of current field theories, theories that will have wider, or simply different invariance groups. What is needed is a physical and intuitive clue pointing the way.

Another investigation is aimed at accomodating general relativity to other physical theories, which nowadays are quantum field theories [7]. Though presently gravitational radiation can be observed only marginally, in the not too distant future it will undoubtedly be possible to observe individual quanta of gravitational radiation. Theoretically, one should be able to predict both the kind of variables that cannot be observed in pairs with arbitrarily high accuracy, and those variables that have discrete spectra. But the elaboration of a quantum theory of gravitation is beset with special difficulties. Some of these are undoubtedly technical, in view of the high degree of non-linearity of EINSTEIN'S field laws. Others appear to be deep-seated. Poincaré-invariant theories dispose of a rigid space-time scaffolding, which, for instance, permits the setting-up of definite Lorentz-invariant propagators. In general relativity this background scaffolding does not exist.

If one accepts the spirit of the theory, there can be no such thing as a classical background metric, though there will be (non-local) constants of the motion. It would seem sensible to reformulate the theory in such a manner that all statements and predictions are independent of the choice of coordinate system. Such an endeavor leads one, almost by compelling force, to the adoption of the field over the four-dimensional manifold, rather than of the manifold by itself, as the physically meaningful object to be considered. At least some of the approaches adopted are in accord with the argument presented previously, to the effect that world points are pale and elusive objects.

In spite of sustained efforts, a definitive and generally accepted quantum theory of gravitation does not as yet exist. Opinions vary as to what features such a theory should possess in order to be universally acceptable. Some serious workers consider the whole program of quantization unsound; they would prefer to retain a classical metric field, which would serve as a backdrop to all non-gravitational physical force fields. The non-gravitational fields are, however, to be quantized. If the gravitational field is to be quantized along with all other fields, opinions differ as to whether such a theory should possess an invariance group of the scope of the group of curvilinear transformations, or whether it should be merely Poincaré-invariant. Finally, it is not at all clear whether the topological aspects of space-time are to be quantized, or whether a quantum state should resemble an ensemble of (classical) metric fields all defined on homeomorphic point sets. Most of the workers in this field, including the present author, have strong but differing opinions on all of these points; the very fact that there is disagreement would seem to indicate that so far these opinions should be regarded as just that, not as universally agreed-to principles.

In the absence of a complete quantum theory of gravitation, one may attempt to obtain some tentative ideas on the scope of indeterminacy relations in the final theory. Guidance is provided by the Poisson brackets of the classical theory. Poisson brackets can be defined between constants of the motion, and presumably approximate brackets between approximate constants of the motion. Admittedly, approximate results are no more to be trusted than the underlying approximation or perturbation approaches. Relatively little actual calculations have been done. One calculation, which was not approximate but based on the notion of observables by intrinsic coordinates (also known as coincidences), has resulted in Poisson brackets between components of the metric tensor so involved that intuitive interpretations were not practicable.

Another interesting question involves the recognition of observables with discrete spectra. Recently I have proposed, for a preliminary survey, a phase integral quantization scheme of the kind that held sway before 1925. To formulate such phase integrals one need not appeal to perturbative ("weak-field") notions but merely to topological properties of certain sub-manifolds of the phase space. This program has not yet progressed far enough that I dare predict its ultimate utility [8, 9].

To summarize, theoretical research in gravitation theory, aside from its astronomical and cosmological implications, which I have not even mentioned in this article, challenges our traditional views of space and time, of point sets, and of geometry underlying our theory-making in the physical sciences, and suggests new departures to those who have a bent for speculating along these lines.

References

1. LEVERRIER, URBAIN: Théorie et tables du mouvement de Mercure. Ann. Obs. Imp. Paris **5** (1859).
2. EINSTEIN, A.: Ann. Physik **49** (1916). Reprinted (in English) in Einstein et al., The principle of relativity, Dover *and* A. Einstein, Preuss. Ak. Wiss. Sitzungsber. 1915, 831–839.
3. TONNELAT, M. A.: Les théories unitaires de l'électromagnetisme et de la gravitation. Paris: Gauthier Villars 1965.
4. EINSTEIN, A., ROSEN, N.: Phys. Rev. **48**, 73–77 (1935).
5. WHEELER, J. A.: Einstein's vision. Berlin-Heidelberg-New York: Springer 1968; – Phys. Rev. **97**, 511 (1955).
6. BERGMANN, P. G., KOMAR, A. B.: To be published.
7. – Helv. Phys. Acta, Suppl. IV, **79** (1956).
8. – To be published by New York Academy of Sciences;
9. – GRG Journal **2**, 37 (1971).

5 Geometry and Physics: A Global Approach

E. J. Post

M.I.N.D. Inc. Cambridge Mass.

(Multidisciplinary Institute for Neuropsychological Development)

Since global and topological notions are common to all geometries, it seems appropriate and relevant to ask about the relationship between topology and physics. In studying the topological aspects of physics one is then, at least temporarily, absolved from the immediate obligation of perhaps prematurely specifying the detailed structure of the geometric substratum underlying physics.

It is certainly not the case that global procedures have permeated contemporary physics. In fact, the literature on the subject is meagre. There is a book by J.A. WHEELER [1] which has done much to initiate this different trend of thinking in physics. This book is much wider in scope than the present article. A justification for the publication of the latter should be seen as based on the discussion of more and other specific examples. Some of the examples (see Section 4 on the MAXWELL equations) have been intentionally presented in a somewhat unorthodox fashion to bring out the contrast with customary local discussions of the subject.

1 The Local and Global Approaches in Physics

It is almost fair to say that the birth of physics as a quantitative theory took place in the days of NEWTON. The formulation of the Newtonian equations of motion were a breakthrough in the field of science. This breakthrough should be classified as local in nature. The equations of motion are a set of differential equations from which trajectories of mass-points or centers of gravity can be deduced, provided initial conditions, constraints and force laws are given.

The abstraction known as the mass-point is all by itself a typical example of a "localizing" of concepts. It thus appears that the first great quantitative breakthrough in physics indeed was a "local" one.

Afterwards, attempts were made to understand the laws of motion from a more global point of view. FERMAT's principle for the minimal length of a light path is a striking example of a global conceptualization of a problem. MAUPERTUIS, EULER, LAGRANGE and finally HAMILTON applied similar thoughts to the motion of point-particles. They thus succeeded in giving at least a partial globalization to the inherently local concepts of point mechanics.

The extension and generalization of the laws of point dynamics to the dynamics of continuous media was largely the single-handed work of EULER. He made a distinction between deformable and rigid bodies. The equations of rigid body dynamics that are used today have his name and were first formulated by him. The equations of fluid dynamics and the fundamental equations of the theory of elasticity also owe their present existence to EULER's formulation of the conservation laws of momentum for the continuous medium. EULER's starting point was the Newtonian point-particle formulation.

It thus appears that also continuum dynamics emerged essentially as a local theory. The global counterparts in the form of a generalized Hamilton principle still play a relatively minor role in modern continuum dynamics. They have not yielded critical new insights that go beyond the point-particle based initial formulation.

A striking example of a fundamental law that got off the ground as a global law is FARADAY's induction law, but soon after its discovery it was converted into a local differential law by MAXWELL (curl $E = -\dot{B}$). By that time the local approach was so well developed that one felt a need and perhaps a relief whenever a global law could be reexpressed in local terms.

The conversion from global to local laws depends on a couple of integral theorems named after STOKES and GAUSS. These theorems are of the greatest importance for a study of the interrelation between global and local approaches in physics. For that reason a special section of this article has been devoted to the so-called generalized STOKES theorem, which contains as corollaries GAUSS' theorem, GREEN's theorem, the original STOKES theorem, as well as the fundamental theorem of the integral calculus.

After this preliminary exploration it is not hard to classify more recent developments in physics according to their local and global aspects. For instance, the gravitational field equations of the general theory of relativity originated as an attempt at formulating a local law. However, the solutions of these equations have, of course, a very pronounced global overtone. In fact, the global behavior or the desirability of enforcing a specific global behavior of these equations have led to modifications of the original equations. These modifications are normally intended to accommodate certain views concerning the cosmological structure of the universe.

It is clear from the previous example that a healthy development requires the possibility of going back and forth between the local and global points of view. However, not all local relations lend themselves to meaningful global extrapolations.

Quantum mechanics is an example of a recent theoretical development which has a much stronger global overtone than its classical counterpart, point mechanics. Quantum mechanics started out as a calculational scheme chiefly concerned with spectral behavior. In its original formulation (Heisenberg) it had no immediate bearing on basic assumptions concerning the geometry of physical space until SCHRÖDINGER reformulated the theory by obtaining it from a variational principle (global overtone). The mathematical condition for obtaining these so-called stationary states are typically global conditions. They are: (1) the uniqueness of the wave function and (2) the square integrability of the wave function over the whole space.

The conditions of square integrality and uniqueness depend, strictly speaking, on a topological specification of physical space. So far, it has been satisfactory to assume an infinitely extended Euclidean space. It is not inconceivable that future application may depend more critically on the details of this topological specification.

It follows from the presented examples that the local and global points of view have nothing to do with the common distinction between micro and macro systems in physics. A microsystem such as the hydrogen atom requires that its wave function be defined throughout space. Conversely, there are macro theories such as continuum dynamics that lack a satisfactory global formulation.

2 *Some Global Concepts in Mathematics* [1]

The word "global" came into prominence as a specific mathematical notion when differential geometry broke away from its historical confines as a mostly "local" discipline. The contrast between the local and global aspects of differential geometry is perhaps best illustrated in reference to the analytic tools that are being used for studying local and global aspects of surfaces or manifolds in general.

The analytic machinery of local differential geometry manifests itself as a collection of differential expressions relating to local geometric properties in the neighborhoods of points of lines, surfaces or manifolds in general. In global "differential" geometry one is by contrast interested

[1] See Refs. [2] through [6].

in the integrals of these local differential relations. Questions arise whether these integrals exist and whether they are unique. If they are not unique, what procedures are available to render them unique? Differential geometry in this sense could be, and is at times, referred to as integral geometry.

When differential geometry emerged as the study of curved lines and surfaces in three-dimensional space, it showed a quite satisfactory balance between local and global results: the Gauss-Bonnet theorem is a beautiful example of a global theorem. The reason for the local-global balance was that it was relatively easy to visualize topological configurations in three-dimensional space.

With the development of differential geometry in higher dimensional manifolds it became increasingly difficult to have the global insights keep up with the local insights. The work of POINCARÉ on the homology classification of polyhedral configurations slowly opened the door to more meaningful global results. These concepts were further developed by CARTAN, DE RHAM and many others. It is relevant for the purposes of physics to concentrate on the contributions of DE RHAM [4].

Intimately tying in with the work of DE RHAM there is the generalization of some classical integral theorems known to every physicist: the theorems of GAUSS, GREEN and STOKES. The generalized version of these theorems contains all three as special cases, even including the fundamental theorem of the integral calculus. This generalized encompassing version of these theorems is normally referred to as just STOKES' theorem. This STOKES' theorem in its general form is concerned with integrals of differential forms. The coefficients of these differential forms, primarily when referred to a cartesian frame, are the field quantities of the physicist. These fields have the transformation properties of completely antisymmetric covariant tensors and their physical meaning can be extended to arbitrary frames of reference. The valence of the tensor (its number of subscripts) corresponds to the degree (number of integration variables) of the differential form. Tensors of this nature play an important role in physics.

The Stokes theorem in its general form is known to be instrumental in relating global and local statements in physics. For instance, the Maxwell equations as local statements are inferred from the corresponding global integral statements through the use of STOKES' theorem. The global statement, e.g. Faraday induction law, normally represents the direct results of an experimental observation. The local statement is inferred through STOKES' law.

A striking analogue of the CAUCHY residue theorem, in the sense that it is related to the multidimensional generalization of STOKES' theorem, was first conjectured by CARTAN. It was subsequently formulated in a complete form and proved by DE RHAM. It is known as DE RHAM'S theorem. The so-called de Rham periods take the place of the Cauchy residues. The de Rham periods, though, do not necessarily share the explosive feature so characteristic of the residue pole of complex functions.

It is instructive to pursue this analogy further in order to obtain a clearer idea of the meaning of the de Rham theorem and its importance for physics. In doing so one needs the local and global analogues of what in complex function theory would be called local and global analyticity.

Analyticity in complex function theory requires that the derivative be independent of the direction along which two points are made to coincide. It follows that the real and imaginary components of the function satisfy the so-called Cauchy-Riemann equations. These are differential equations which inform us about the local analyticity in the neighborhood of a point. The local condition of analyticity can be extended into a global condition by simply requiring that the Cauchy-Riemann equations are met in a domain of definition of the function. The function is then called holomorphic or regular in that domain.

Global analyticity can also be approached in a manner that is more global in nature. It can be defined by demanding that certain line integrals along arbitrary contours inside a domain vanish; to define these contour integrals uniquely it is necessary that the functions be single valued. This method of defining global analyticity carries over better when defining the corresponding notions for fields.

For a field a condition corresponding to the Cauchy-Riemann equations is that one requires that the exterior derivative of a field vanishes. The field is called "closed" in the neighborhood of a point where its exterior derivative vanishes. Hence, the property of a field of being closed is a local property. A corresponding global property cannot now be simply obtained by requiring that a field be closed everywhere in a region. The reader is reminded that the word "field", as it is used here, can be replaced by the expression "differential form".

It may be appropriate to mention that the familiar operations of curl and divergence are special cases of the exterior derivative—which should not be construed to mean that any curl and divergence operation in physics can be meaningfully translated into an exterior derivative operation.

The distinction between the concepts of closed and analytic shows up if we examine the condition for a two-dimensional case. Closed is then only the Cauchy half of the Cauchy-Riemann equations. The analogue of the Riemann half requires that the field be also *co-closed.*

To define the operation co-closed requires that the manifold in which the field is defined has a metric. The complex plane has a builtin metric: the modulus of a complex quantity. It follows that co-closed is not an intrinsic topological property of a field, but rather gives a mutual porperty of field and metric.

In keeping with the earlier adopted philosophy of concentrating on intrinsic topological properties of fields, one would like to avoid an untimely commitment to a specific geometric structure such as a metric structure. Hence, one would like to avoid the operation associated with co-closed.

It is interesting to know that the de Rham theorem does not require the notion of co-closed. This fact enhances the intrinsic topological nature of the theorem and illustrates an important distinction between the Cauchy and de Rham theorems.

If one now raises the question how to obtain the global counterpart of the local concept of a field being closed, one may recall that global analyticity could also be defined by requiring that certain contour integrals in a region vanish.

Similarly, one obtains a global extension of the local concept *closed* if one requires that all "cyclic" integrals of a closed form vanish. A form that satisfies this criterion is called *exact*. The term "cyclic" denotes that the integrations are performed over *closed* domains in the region. A boundary of a region is closed, but any closed domain in a region is not necessarily a boundary. The sum of a number of closed domains may constitute a boundary. Example: an annular region is bounded by two closed (cyclic) domains, which together form a boundary, but not individually. A somewhat imprecise and unconventional wording of the de Rham theory may now be given in the following manner.

The first part of the de Rham theorems claims that a closed form is exact if all its cyclic integrals vanish. The form (field) may then be regarded as the exterior derivative of a form (field) of lower degree (lower covariant valence).

The second part of the de Rham theorem makes a statement about the case where not all cyclic integrals of a closed differential form vanish. These nonvanishing cyclic integrals are called periods. The number of independent periods equals the rank (Betti number) of the homology group characterizing the topological structure of the domain of integration.

Two periods are called independent if the cyclic domains of integration together don't constitute a boundary.

It is now useful to conclude with an explicit formulation of the Stokes and de Rham theorems in terms of two different mathematical shorthands.

Stokes' Theorem. Let M be a $(k+1)$ dimensional, compact, oriented manifold with boundary. The boundary is closed of dimension k and is denoted by ∂M. Let F (an infinitesimal) be a regular k-form (i.e. of degree k) defined over M. The exterior derivative of F be denoted by dF. The theorem is then expressed by the relation

$$\int_M dF = \int_{\partial M} F. \tag{1}$$

Eq. (1) renders STOKES' theorem in a very austere form, yet it covers all the conceptual essentials. Note how the boundary operator ∂ and the (exterior) derivative operator d exchange places in Eq. (1). These operators share the Property $\partial\partial=0$ and $dd=0$, because the boundary of a boundary vanishes as well as the alternation of partial derivatives in d.

A rendering geared to calculational purposes requires a coordinate version of (1) which can be advantageously presented in a tensorial short-hand.

Let M be imbedded in a manifold N of dimension $n \geqq (k+1)$. Let there be a single coordinate neighborhood on N (with coordinate labels $\lambda, \nu = 1, 2, \ldots, n$), completely covering M and its boundary ∂M.

The differential form for F, for $k=2$ say, is then given by

$$F = F_{\lambda\nu}\, df^{\lambda\nu}. \tag{2}$$

Its exterior derivative is

$$dF = \partial_{[\kappa} F_{\lambda\nu]}\, df^{\kappa\lambda\nu} \tag{3}$$

where $[\kappa\lambda\nu]$ denotes an alternating sum, which now makes the property $dd=0$ evident, because $\partial_{[\kappa}\partial_{\lambda]} \equiv 0$ when operating on well-behaved functions. The reason for denoting the differential form by F rather than dF also manifests itself in Eq. (3). The integration elements of M and ∂M are denoted by $df^{\kappa\lambda\nu}$ and $df^{\lambda\nu}$. They are completely antisymmetric in the indices and can be expressed through parameter lines $q_1, q_2, \ldots,$ which can be aligned, if convenient, with the coordinate neighborhood on N

$$df^{\lambda\nu} = dq_1^{[\lambda}\, dq_2^{\nu]} \quad \text{etc.} \tag{4}$$

The Stokes theorem now reads (for $k=2$)

$$\int_M \partial_{[\kappa} F_{\lambda\nu]}\, df^{\kappa\lambda\nu} = \int_{\partial M} F_{\lambda\nu}\, df^{\lambda\nu}. \tag{1a}$$

The extension to other values of k follows exactly the same pattern. Note how the calculational version requires many detailed specifications in comparison with the austere conceptual version given by Eq. (1) notwithstanding the use of the rather efficient short-hand of tensor calculus. However, it would be difficult to dispense completely with these cumbersome calculational procedures, particularly in physics.

de Rham's Theorem. Let the differential k-form be closed; $dF=0$. Let $Z_l(k)$ denote a family of all cyclic domains of integration of dimension k, hence $\partial Z_l(k)=0$ for all l.

The integrals

$$\int_{Z_l} F=\alpha(l)$$

are then called periods of F. Parts I and II of the theorem are:

(I) If $\alpha(l)=0$ for all l, F is called exact and is the exterior derivative of a $(k-1)$ form A

$$F=dA.$$

The cyclic domains $Z_l(k)$ are boundaries of $(k+1)$ dimensional domains if the periods of all closed F vanish.

(II) The form F is not exact if $\alpha(l)\neq 0$ for some l. The form F is then determined, modulo an exact form, if all of its independent periods are given. The number of independent periods equals the number of infinite cyclic generators of the homology group H_k characterizing the topology of the k dimensional domains of integration (see Section 3).

A coordinate rendering of this theorem, would, of course, follow the same pattern as for the Stokes theorem.

3 *Physical "Obstruction" Theory (Homology and Homotopy)* [1]

POINCARÉ, who initiated the homology method for classifying the topology of geometric configurations, also conceived of a procedure which in many ways is complementary to homology. The procedure is known as the homotopy method. It is a classification based on the contractibility by deformation of cyclic domains of integration around "obstructions". On the basis of the latter feature, one may sense already a connection with the de Rham theorem.

By contrast, the homology classification has been constructed around a group structure based on the additive feature of integration. This additivity permits a characterization by means of an additive abelian group. The cyclic domains of integration then must be characterized by an abelian group that has cyclic elements. Let the group characterizing the totality of cyclic domains of dimension P be denoted by Z_p. Since not all cyclic domains have the properties of a boundary, one may expect the group Z_p to have a sub-group B_p characterizing only the boundaries. It is difficult to give a unique realization of the groups Z_p and B_p because the process of integration is based on a limiting process of subdivision which depends on the definition of integration (e.g. RIEMANN-LEBESGUE). However, POINCARÉ discovered that the factor

[1] See Refs. [7] and [8].

group $H_p = Z_p/B_p$ has a unique and invariant meaning for the topology of the geometric configuration of the integration domain that is being considered. The group H_p is called the $p\,th$ homology group of the configuration. POINCARÉ obtained this result by approximating the integration domains by a finite process of "simplicial" subdivision. He was able to show that the group H_p is independent of this process of subdivision.

While the group structure of homology is based on a specification of the integration process itself, the group structure of homotopy is based on a specification of families of equivalent cyclic (closed) integration domains. The set of all possible integration domains of given dimension p is subdivided into equivalence classes of domains that can be deformed into one another. Domains belonging to the same class are called homotopic. The subdivision into distinct equivalence classes comes about by the rule that deformation are not permitted to go through a so-called obstruction. Obstructions are obstacles to be respected by the deformation game. The pole of a complex function is an example of an obstruction. A contour (cyclic domain) enclosing the pole cannot be contracted to a point without going through the pole.

For 1-dimensional contours one can now define a group operation on the set of equivalence classes by considering two subsequent operations over contours belonging to different classes. The group thus obtained is called the fundamental group. The fundamental group is not in general abelian. A commuting factor group of the fundamental group exhibits a simple algebraic relation with the earlier defined homology group H_1.

The number of independent cyclic elements of infinite order of a homology group is the rank of the group and equals the number of independent periods mentioned in DE RHAM'S theorem. This holds for 1 as well as for higher dimensional domains (H_p, $p > 1$) of integration.

It follows from this summarized presentation of homology and homotopy that "obstructions" are of the very substance of topology. To evaluate the importance of obstruction theory for physics, one has to ask what entity in physics constitutes an obstruction to another entity. The answer tends to be trivial but is nontheless revealing from a point of view of ordering concepts.

It can hardly be denied that the concept of a purely geometric obstruction has been generated by man's ability to perceive the absence and presence of matter in physical space. To abstract this notion of obstruction so that it is independent of the distribution of matter is a small step in terms of imagination, but a difficult one to carry through when the visualization breaks down. That was the very reason for POINCARÉ'S ingenuous but tedious system of homology and homotopy classification.

Starting from the abstract obstruction concept that was induced by the physically crude and primitive distinction presented by the distribution of matter in space, one can somewhat sense the ramifications that are conceivable from a point of view of physics. The crudely collective concept of matter can be resolved into a number of much more sophisticated physical concepts related to the structure of matter.

The major task of physical obstruction theory should be the identification of truly basic physical objects of obstruction. Their dimensionality in the spacetime manifold should be established, as well as their subsequent relation to fundamental physical laws.

In the light of the preceeding discussions it seems useful to think of an obstruction as something inside a cyclic domain of integration in the sense of the de Rham theorem. An obstruction is then revealed by the existence of nonvanishing de Rham periods. Conversely, if all de Rham periods vanish, one may conclude that the suspected obstruction does not exist (e.g. magnetic charge and the creation or annihilation of net electric charge).

Following is a table of some fundamentally important cyclic integrals in physics with vanishing and nonvanishing periods:

Table 1

Cyclic domains of integration		Nonvanishing de Rham periods	Physical dimension	Vanishing de Rham periods	Physical dimension
Dimensions of domain of integration	1	Angular Momentum	$\hbar$	—	—
		Magnetic flux	$\hbar/e$	—	—
	2	Electric charge	e	Absence of magnetic charge	$\hbar/e$
	3	Action	$\hbar$	No creation of net charge	e

The symbols $\hbar$ and e are meant to indicate the physical dimensions of PLANCK's quantum and the electronic quantum of charge respectively.

In the next section we will examine the differential forms (fields) associated with these cyclic integrals.

4 *Local Relations with Global Conscience*

It is probably realistic to note, even if one improves the global methods and insights in physics, that the local approach will still prevail by far over the global approach; the reason being that the local approach is used to synthesize arbitrary global situations. The benefits to be expected from studying independently the global aspects of physics, however, would be greatly enhanced if one could retain, more than is presently customary, the global features of local relations. Such a change in policy would facilitate the going back and forth between global and local expressions. In cases where the global formulation is lacking, it could alert us to a better recognition of the specific features that are missing to obtain satisfactory global formulations.

From the preceding discussions it may be clear that, in particular, local physical relations that can be cast into a differential form are eligible for subsequent globalization via the Stokes and the de Rham theorems. It may come as a surprise that one of the first applications of the de Rham theorem in physics was discussed by EINSTEIN a considerable time prior to its definitive formulation.

EINSTEIN's motivation for using a topological consideration was prompted by an effort to find an invariant formulation of the so-called Bohr-Sommerfeld quantization conditions. The Bohr-Sommerfeld method has been superseded by the formulation of modern quantum mechanics but, in an asymptotic sense, it still retains a meaning in the modern theory. Let us reexamine EINSTEIN's analysis while using more freely the notions and terminology of modern mathematics.

The motion of a particle according to the Hamilton-Jacobi theory is governed by the integrability of the differential form of action W:

$$dW = \sum_{l=1}^{3} p_l \, dq^l - H \, dt. \tag{1}$$

The Hamiltonian H contains all the information about the physical environment in which the particle moves. By consulting any appropriate text on analytic dynamics, one finds that the Hamilton equations of motion make the form integrable so that one may say that the four-momentum is locally a gradient:

$$p_l = \frac{\partial W}{\partial q^l} \quad \text{and} \quad -H = \frac{\partial W}{\partial t}. \tag{2}$$

This gradient property cannot hold in a global fashion, because there are closed loop integrals that do not vanish, e.g.

$$n_\varphi h = \oint p_\varphi \, d\varphi = 0 \quad \text{and} \quad n_r h = \oint p_r \, dr. \tag{3}$$

φ is the azimuthal and r is the radial coordinate, p_φ and p_r are the conjugate momenta, and n_φ and n_r are the corresponding quantum numbers.

The compatibility of (2) and (3) implies that the Stokes theorem cannot be applied, because the cyclic domains of the integrals (3) are not boundaries. It also follows that the relations (2) can only have a *local* validity. In other words, the differential form (1) is closed; it is not exact. That means the exterior derivative of (1) vanishes. It is not true that the four-momentum $(p_l, -H)$ may be regarded in a global fashion as the exterior derivative of W. (Note that the exterior derivative of $(p_l, -H)$ is equivalent to a four-dimensional curl of the momentum four-vector.)

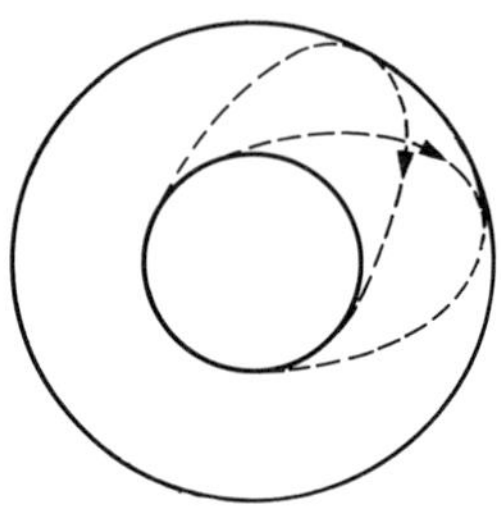

Fig. 1. Family of orbits for a planetary motion of given energy and angular momentum

What is the obstruction that prevents the cyclic integration loops of the integrals (3) from being contracted to points? Equivalently: why are the cyclic domains of the integrals (3) not boundaries?

To answer these questions, let us consider a planetary type motion such as illustrated by the naive model of the hydrogen atom. For a given energy and angular momentum of the system, the possible orbits are ellipses contained in an annular region (see Fig. 1). The corresponding momentum field p is likewise contained in this region. Note that this momentum field is two-valued. One selects a branch of the p field according to whether one travels from the inner to the outer envelope or vice-versa (Fig. 1). For the integration of a two-valued momentum field, it is necessary to specify the branches that have to be considered when completing a path of integration. An examination of Fig. 1 shows that one switches branches at the outer and inner envelope of the family of orbits. Following Einstein one can make the momentum field single-valued if one separates the two p-branches by defining them to exist on two distinct sheets that are connected along the envelopes. Fig. 2 illustrates how the annular region of Fig. 1 becomes a torus. An actual orbit travels for half of its length on the upper sheet Q and for the other half on the lower sheet P. It wraps itself once around the torus (meridionally) when completing an orbit.

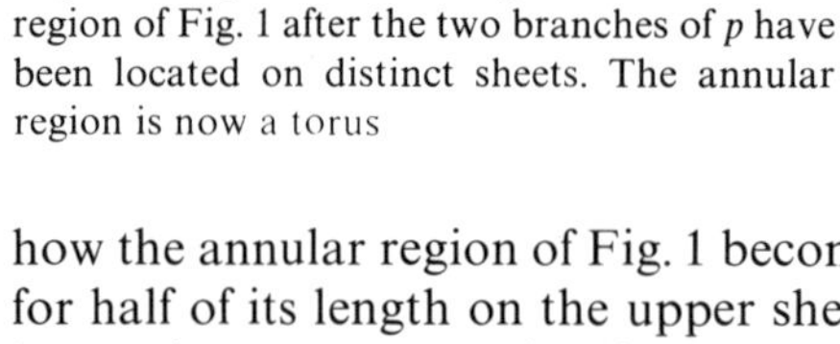

Fig. 2. Perspective drawing of the annular region of Fig. 1 after the two branches of p have been located on distinct sheets. The annular region is now a torus

What is the physical obstruction? The answer is that only the surface of the torus is accessible to particles in orbits of the given energy and angular momentum; hence, the rest of space inside and outside the torus is forbidden orbital territory and constitutes an obstruction.

Now one can easily prove with the help of STOKES' theorem that all closed loop integrals of p along a meridional section have the same value, because two meridional loops together form a boundary (condition for the applicability of STOKES' theorem) of a region on the torus inside of which curl $p=0$ (e.g. the meridians going through P and Q in Fig. 2). A similar argument holds for arbitrary equatorial loops which do not wrap themselves around the torus.

It follows that EINSTEIN's construction for making the momentum field regular (i.e. single valued) leads to two topological invariants; they are the quantum conditions of Eq. (3), which we may now recognize as de Rham periods. There are not more than two independent periods. An actual orbit is easily seen to be a combination of two periods, an equatorial and a meridional period. These two periods correspond to the two infinite generators of POINCARÉ's homology group H_1 characterizing the topological structure of the torus.

Mathematicians do not normally think of EINSTEIN as a mathematician which is probably fair and correct. However, this rather unknown article, which was prompted by EINSTEIN's urge as a physicist to find an invariant characterization of the Bohr-Sommerfeld quantum conditions, illustrates beautifully how a physicist can come close to formulating a major theorem in topology and analysis. Indeed, many of the ingredients that go into the de Rham theorem were present in EINSTEIN's article.

The obstruction associated with orbital motion is as abstract an obstruction as one may encounter in a context of physics. In fact, it is more a mathematical abstraction than a physical obstruction, but so is the pointparticle itself whose orbital behavior is being assessed. Which goes to show that one abstraction may lead to another, a state of affairs acceptable to mathematics. In physics, however, one likes to impose additional constraints on the art of making abstractions. So let us direct our attention to the more explicit physical obstructions, such as electric charge or magnetic flux.

Proceeding in this vein one can make the observation that the conservation of flux means the absence of magnetic monopole sources. Similarly, conservation of charge means the absence of a mechanism for the creation of net charge; whenever charge is created it is always in equal amounts of opposite polarity.

The conservation of flux is expressed by div $B=0$, where B is the magnetic induction. It is easy to cast this statement into the garb of a differential form. It says that the exterior derivative of the 2-form

$$B_{lk}\, d\underset{1}{x}^l\, d\underset{2}{x}^k, \qquad B_{lk}=-B_{kl}$$

vanishes: $\partial_{[s}B_{lk]}=0, \qquad l,k=1,2,3$

symbolically: $dB \quad =0.$

The vector B (or B^l) is related to the antisymmetric tensor B_{lk} by means of LEVI-CIVITA's alternating symbol for three-dimensions.

The statement $dB=0$ expresses the local absence of magnetic monopole sources (magnetic charge). This local statement becomes a universal global statement if all integrals $\int B$ over cyclic domains vanish. Then there exists, according to DE RHAM's theorem, a one-form A (vector potential) with the property $B=dA$.

If $B=dA$ expresses that there exists no magnetic charge in all of physical space, how do we extend this statement to space-time? If one were to succeed, one might then say at no time does magnetic charge exist anywhere in physical space. It is well known that this extension requires the definition of an additional spatial field, the electric field E. The fields E and B are known to combine into an antisymmetric tensor $F_{\lambda\nu}$, which defines a differential 2-form over space-time.

$$F=F_{\lambda\nu}\,df^{\lambda\nu}, \qquad \lambda, \nu=0, 1, 2, 3. \tag{4}$$

The vanishing of integrals of F over all cyclic domains implies, according to DE RHAM's theorem, the existence of a one-form from which F can be derived by exterior derivation. If we take for this one-form the same symbol A as previously used for the spatial components of the vector potential, one has

$$\text{for all cyclic domains} \quad \int F=0, \tag{5}$$

$$\text{hence, } F \text{ is exact and} \qquad F=dA \tag{6}$$

where A is defined modulo a closed form—the gauge property of A. For a global statement, it would be better to say that A is defined modulo an exact form.

It follows from (6) that

$$dF=0 \tag{7}$$

which, in vector language, means

$$\operatorname{div} B=0, \qquad \operatorname{curl} E=-\dot{B}. \tag{7a}$$

The second Eq. (7a) is the differential form of FARADAY's induction law, which ensues here as a corollary of demanding the general absence of magnetic charge for all past and future times anywhere in physical space. Hence, the Eq. (7) extends the purely spatial result $dB=0$ into an invariant statement for space-time.

The absence of a mechanism for creating net charge for all past and future times anywhere in physical space can be treated in a similar manner. The conservation of net charge is normally given by the familiar equation

$$\operatorname{div} S+\frac{\partial \rho}{\partial t}=0 \tag{8}$$

in which S is the current density and ρ the charge density.

In a space-time symmetric version one may write

$$\partial_\lambda S^\lambda=0, \qquad \lambda=0,1,2,3. \tag{8a}$$

This four-divergence can be converted into an exterior derivative of a 3-form by means LEVI-CIVITA's alternating symbol for four-dimensions:

$$E_{\lambda\nu\sigma\kappa}$$
$$S^\lambda E_{\lambda\nu\sigma\kappa}=S_{\nu\sigma\kappa}.$$

The corresponding differential form S is

$$S=S_{\nu\sigma\kappa}\,df^{\nu\sigma\kappa}.$$

The local conservation of net charge is then expressed by

$$dS=0. \tag{8b}$$

If all 3-dimensional integrals of S vanish

$$\int S=0 \tag{9}$$

for all cyclic integration domains, then it follows from (9) that S is not only closed (8b) but also exact.

It follows that there exists a 2-form G such that

$$dG=S. \tag{10}$$

The two-form G is defined modulo an exact form. (Gauge property of G.)

The Eq. (10) in customary vector language reads

$$\operatorname{div} D=\rho, \qquad \operatorname{curl} H=\dot{D}+S \tag{10a}$$

where D and H are two spatial vectors which together constitute a space-time 6-vector according to

$$G=G_{\lambda\nu}\,df^{\lambda\nu}. \tag{11}$$

Note the distinction with the differential form F, where F is exact and does not exhibit a gauge property: G is not exact, not even closed, and does exhibit a gauge property. In other words, whereas F is precisely defined as a physical quantity, G is only defined modulo an exact 2-form – a property it shares with the one-form of the vector potential A.

The gauge ambiguity of the vector potential A is a generally accepted feature of electromagnetic theory, not so the gauge ambiguity of G, i.e., D and H. It is generally agreed that the quantities D and H are subject to serious problems of physical interpretation. It is not customary, however, to relate these difficulties to a mathematical gauge property, such as is readily accepted for the vector potential A. It may be seen as a definite advantage of the global approach that the deeper mathematical reasons for the interpretation difficulties encountered with D and H can be uncovered.

The presently prevailing attitude in physics is that one is willing to consider only D and H as realistic physical entities if they can be derived from microphysical properties of the medium. This procedure is essentially a solid-state physics problem and not a point of major concern for electromagnetic theory. The modern trend in E.M. textbook writing is accordingly to emphasize the field quantities E and B and to circumvent and evade as much as possible the quantities D and H unless they can be clearly interpreted in a microphysical sense.

If we look, from the point of view of a global approach, at this modern trend in textbook writing, then it becomes clear that this development is based on inadequate information. One can respect the honesty that leads to the decision of eliminating from the scene those quantities that have no well-defined meaning. A gauge property is, however, a well defined property – provided it is recognized as such.

The vector potential is an accepted concept, notwithstanding its gauge property, because here there was never any doubt that such a property existed. In the same vein, the D and H field should be retained as proper physical quantities provided we acknowledge their gauge property.

However, the vector potential should be abandoned if one entertains the idea of admitting the magnetic charge as a physical reality, because then F would have non-vanishing periods and the first part of de Rham's theorem would not apply. Similarly, the D and H fields should be abandoned if and only if one entertains the hypothesis of net charge creation. As long as there is neither proof of one nor the other, we do well to retain D and H along with A, at least in the textbook literature.

The objection against D and H in the modern textbook treatment is, however, never based on a breakdown of net charge conservation. The conservation of electric charge is invariably stated as an indisputable truth, yet there is that undertone of a suggestion that it would have been much better if the older textbooks had not made so much fuss about the introduction of D and H. It seems that a global approach can easily absorb objections of this nature.

The absence of magnetic charge and the absence of net charge creation have a postulational quality in common with the second law of thermodynamics. They are negations of conceivable observations. Their validity has only been established in the sense that so far exceptions have not been observed. The understanding of the nature of the second law statement of thermodynamics has been much improved by statistical mechanics. It is conceivable that the understanding of the other two statements might be much improved by quantum mechanical considerations. For special studies one should certainly accept an analysis of the consequences of having magnetic charge and net charge creation. It should be helpful, though, if one not only considers the local but also the global ramifications of such assumptions.

After this plea for retaining D and H, it is natural to ask: What do we do with the gauge? In analogy with the comparable situation for the vector potential A, one would seek to resolve this matter by establishing an appropriate gauge for D and H according to need and convenience and compatible with the specific problem under consideration. It is this type of discussion that tends to make textbooks lengthy. For the purpose of this article a number of general remarks may serve better the goal of permeating our thinking in physics with global and topological features.

The customary textbook introduction of the Eqs. (10a) is based on COULOMB'S, AMPÈRE'S and BIOT-SAVART'S laws of force interaction between currents and magnets. MAXWELL'S displacement current then follows from a compatibility requirement with the conservation of charge. Note that this reasoning seeks a mathematical formulation of force laws. Modern textbook writers thus have a good point in preferring the force fields E and B in the Eqs. (10a) rather than the fields D and H.

By contrast, the Eq. (10) obtained through the global approach are a consequence of charge conservation. Hence, the global approach enables us to bring to fruition MAXWELL'S original argument for obtaining the displacement current; not only the displacement current but the full formula follows from charge conservation. However, force laws do not play a role in deriving Eq. (10). It follows that the fields D and H can only be informative about a current action of some sort. There is as yet no information about a force interaction associated with current and charge. The dimensional properties

$$[D]=[e\,l^{-2}] \quad \text{and} \quad [H]=[e\,l^{-1}\,t^{-1}]$$

on the one hand and

$$[E]=\left[\frac{\hbar}{e}\,l^{-1}\,t^{-1}\right] \quad \text{and} \quad [B]=\left[\frac{\hbar}{e}\,l^{-2}\right]$$

on the other hand make this distinction evident provided a coherent system of units is being used.

In fact, the dimensions of the differential forms F and G convey this information even more concisely

$$[F]=\left[\frac{\hbar}{e}\right], \qquad [G]=[e].$$

The dimension of G depends solely on the dimension of the electric charge $[e]$, while F depends on the ratio of action $[\hbar]=[m\,l^2\,t^{-1}]$ and electric charge $[e]$.

To take the step from Eqs. (10) to (10a) so that the Eqs. (10a) can be related to the force laws of COULOMB, AMPÈRE and BIOT-SAVART, it is now necessary to establish a relation between D, H and E, B. These relations are known as the *constitutive equations* of the medium. For free space they take the well known form

$$D=\varepsilon_0\,E \quad \text{and} \quad B=\mu_0\,H \tag{12}$$

where ε_0 and μ_0 are quantities that interrelate charge and force units. A mixed unit system is based on an assumption $\varepsilon_0=\mu_0=1$. For reasons already mentioned, we abstain from using a mixed unit system.

A further justification for the particularly simple relations (12) can be based on the fact that they are Lorentz invariant relations if the light velocity c is related to ε_0 and μ_0 according to the well known relation

$$c=1/\sqrt{\varepsilon_0\,\mu_0}\,. \tag{13}$$

In fact, one finds that the invariance of (12) even defines an overgroup of the Lorentz group: the conformal group.[1] This observation is due to BATEMAN [10], who established this invariance directly for the free-space Maxwell equations expressed in the mixed unit system. Hence, by using a global approach, one can pinpoint more precisely whence this conformal invariance as well as the Lorentz invariance come: namely, the constitutive Eqs. (12). Because the Maxwell equations in the differential form (7) and (10) are, by contrast, invariant under a general diffeomorphism which includes locally the 16-parameter general linear group.

The latter observation is due to CARTAN [11] but appears in many different forms in the work of WEYL, VAN DANTZIG and others. Hence,

[1] The relation (12) remains invariant under a 6-dimensional orthogonal group of the vector space defined by E, B and D, H. This 6-dimensional group has 15 independent parameters and is isomorphic with a 15 parameter 4-dimensional conformal space-time group.

all symmetry restrictions follow from the constitutive equations, not from the Maxwell equations. This fact should be hailed as a healthy and logical consequence of the global approach, because the constitutive equations and not the Maxwell equations should specify the symmetry properties of a medium.

For a more general medium than the free-space medium described by the relations (12), for instance a material medium, and in a more general situation where this material medium, the frame or both are moving or accelerated with respect to inertial space, one may expect the relations (12) to become much more complicated. Some of these situations have been elaborately discussed in the literature so that there is no special reason to deal with them in detail here.

For a linear and nondispersive medium, in any of these general situations, it still holds true that the total behavior can be represented by a linear map of the coefficients of F into the coefficients of G. The map χ: $F \to G$ converts an exact 2-form F into a 2-form G that is only closed in a nonconducting medium. The 2-forms F and G are of opposite "parity", which means that for a centro-symmetric situation they will not, in general, lead to a relation between D and B or H and E. Such relations do occur in moving and in accelerated media, as well as in a stationary optically active medium.

So far little work has been done on the algebraic and differential comitants of the map χ – let alone on the physical identification of these quantities. The interested reader may verify that the so-called free space impedance $\sqrt{\mu_0/\varepsilon_0}$ is not only a Lorentz invariant. In fact, one would expect it to be that; it also is an invariant under general transformations (diffeomorphisms). In physics one normally expresses this by saying that $\sqrt{\mu_0/\varepsilon_0}$ is a "general relativistic" invariant. It then follows that the fine structure constant $\alpha = e^2/\hbar c$ (mixed units) is a general relativistic invariant, because $\alpha = (e^2/4\pi\hbar)\sqrt{\mu_0/\varepsilon_0}$ (coherent units) is now a product of general invariants. Recent astronomical observations [12] of the spectra of quasars suggest that α is not only invariant, it appears to be a constant throughout the visible universe.

To summarize the results of this section on local relations with global conscience, the following may be said. It appears that a transcription of fundamental physical relations in terms of differential forms is a most helpful expedient for alerting the mind to the global aspects of theory. One may not expect this transcription to apply everywhere in physics. Classical electromagnetic theory is perhaps an exclusive example where this policy can be worked out in an almost complete detail. Table 2 may help to illustrate further the meaning of this statement.

Table 2

Physical identification of differential forms	Physical dimension	Degree of differential forms				Pair	Im-pair	Closed	Exact	General remarks
		1	2	3	4					
Four-momentum p	$\hbar$	×					×	×		If $A \neq 0, e \neq 0$; $p_\lambda = m\dot{q}_\lambda - e A_\lambda$ ($\lambda = 0, 1, 2, 3$)
Four-potential A	$\hbar/e$	×					×	–	–	A can be closed inside a superconductor
E, B-field F	$\hbar/e$		×				×		×	$F = dA$, hence $dF = 0$ ↗ $\operatorname{div} B = 0$ ↘ $\operatorname{curl} E = -\dot{B}$
D, H-field G	e		×			×		–	–	$dG = s$ ↗ $\operatorname{div} D = \rho$ ↘ $\operatorname{curl} H = \dot{D} + s$ G is closed in a charge-free regions
Four-current density s	e			×		×			×	$ds = 0$ local conservation of charge $\operatorname{div} s + \frac{\partial \rho}{\partial t} = 0$
Derived forms: $A \wedge G = N$	$\hbar$			×			×	–	–	N = Action potential. $dN = L$
Derived forms: $A \wedge S = L_1$	$\hbar$				×		×	×		Mechanical component of action
Derived forms: $F \wedge G = L_2$	$\hbar$				×		×	×		E. M. component of action Can vanish inside a superconductor
Derived forms: $L_1 + L_2 = L$	$\hbar$				×		×		×	Sum total of action is conserved ($L = dN$)

The physical dimension of action is denoted by $[\hbar] = [m l^2 t^{-1}]$. The physical dimension of electric charge is denoted by $[e]$. The definitions of "pair" and "impair" follow de Rham-Variétés Differentiables, Ch. II; pair × pair = pair; impair × impair = pair; impair × pair = impair. A form F is closed if $dF = 0$. A form F is exact if there is an A such that $F = dA$. The symbol $\wedge$ denotes a wedge (alternating) product $F \wedge G \to F_{[\lambda\nu} G_{\sigma\kappa]}$. Four-forms are automatically closed. The forms A and G are defined modulo exact forms (Gauge).

Total action being "conserved" ($L = dN$), the exchange between L_1 and L_2 is governed by a (variational) extremum adjustment $\delta \int L_1 = 0$ and $\delta \int L_2 = 0$.

The constitutive properties of a charge-free medium are given by a map $\chi: F \to G$. In vacuum χ reduces to an (anti)diagonal matrix for the family of inertial frames. In superconductors the London equations give the constitutive behavior, they are equivalent to the condition $dp = 0$ for the charge carriers of the supercurrent. It is customary to relate F and G by the vacuum relations in conductors as well as in superconductors.

5 Conclusion

Much conceptualization in contemporary physics is bogged down by unnecessary assumptions concerning a specific choice of coordinate neighborhoods. A more topological way of thinking might resolve some difficulties ensuing from a premature coordinatization. Contemporary physics suffers in this respect of two important syndromes: the cartesian syndrome and the inertial frame syndrome.

Physicists cannot think without a cartesian frame.

Proof. Even if a situation requires the use of curvilinear coordinates, the physicist will faithfully erect at every point a cartesian frame of unit vectors for reference. It is not wrong, but it is unnecessary and, above all, it is symptomatic of a state of mind. A token training in topology has not been able to change that state of affairs.

Secondly, physicists cannot think without explicit reference to an inertial frame.

Proof. Even today the usual derivation of the fundamental equations of fluid dynamics uses inertial frames that move with the fluid but don't accelerate with the fluid. This shows how our conceptualization becomes nonfunctional outside the realm of the inertial frame.

It may be necessary that we mend our ways in physics. In fact, a topological analysis and a possible critique of the process of conceptualization in physics should be regarded as a needed and long overdue purgatory for the existing body of knowledge in theoretical physics.

The simple examples presented in the last section may hopefully serve as an incentive to pursue further this course of action. I may only mention the delightful opportunity we have in E. M. theory to dispense with both syndromes simultaneously, the cartesian as well as the inertial syndrome.

Acknowledgment. Professor A. COPELAND (UNH) and Professor A. NIJENHUIS (Penn.) were kind enough to criticize the mathematical sections of this chapter. The remaining inadequacies, which result mainly from an overcondensed presentation of an important branch of mathematics are, of course, the author's responsibility.

References

1. WHEELER, J. A.: Geometrodynamics. New York: Ac. Press 1962.
2. SPIVAK, M.: Calculus on manifolds. New York: Benjamin 1965. A monograph on the generalizations of Stokes' theorem.
3. FLANDERS, H.: Differential forms. New York: Ac. Press 1963. A discussion of Stokes and de Rham theorems and their relation to differential forms.

4. Rham, G. de: Enseignement Math. **35**, 213 (1936); – Math. Seminar Un. Hamburg **12**, 313 (1936). An older but very readable reference on de Rham's theorem, not leaning too heavily on modern abstract methods: A later more abstract treatment, Variétés Différentiables. Paris: Hermann 1955.
5. Hodge, W.V.D.: Harmonic integrals. Cambridge: Cambridge Un. Press 1952. A very concise discussion of Stokes and de Rham theorems and their relations to homology.
6. Schouten, J.A.: Ricci calculus, 2nd, ed. Berlin-Göttingen-Heidelberg: Springer 1954; – Tensor analysis for physicists. Oxford Un. Press 1951.
7. Hilton, P.J., Wylie, S.: Homology theory. Cambridge: Cambridge Un. Press 1967.
8. Lefschetz, S.: Introduction to topology. Princeton, N.J.: Princeton Un. Press 1949.
9. Einstein, A.: Verhandl. Dtsch. Physik. Ges. **19**, 82 (1917).
10. Bateman, H.: Proc. London Math. Soc. **8**, 77 (1910).
11. Cartan, E.: Ann. Ecole Norm. **41**, 1 (1924). Cartan's differential form discussion of the Maxwell equations differs from the so-called standard general-relativistic treatment by virtue of its being independent of metric structure. A tensorial discussion of this same feature of "natural invariance" of the Maxwell equations is implicit in H. Weyl's treatment (Space-Time-Matter, New York: Dover reprint (1951), first edit. (1918).
In the following references one finds an explicit treatment of the feature of "natural invariance" by tensorial methods:
Dantzig, D. van: Proc. Cambridge Philos. Soc. **30**, 421 (1934); – Proc. Amsterdam Acad. **37**, 521, 526, 644, 825 (1934).
Post, E.J.: Formal structure of electromagnetics. Amsterdam: North Holland 1962.
Bunge, M., ed.: Delaware seminar foundations of physics, p. 102. Berlin-Heidelberg-New York: Springer 1967.
Recent applications of natural invariance to accelerated E. M. systems:
Post, E.J.: Rev. Modern Phys. **39**, 475 (1967).
Anderson, J.L., Ryon, J.W.: Phys. Rev. **181**, 1765 (1969).
12. Bahcall, J.N., Schmidt, M.: Phys. Rev. Letters **19**, 1294 (1967).

IV. Of Quanta

6 Probability in Quantum Mechanics and in Other Statistical Theories [1]

ARTHUR FINE

Department of Philosophy, Cornell University
Ithaca, N. Y.

1 Introduction

In the ongoing investigation of the conceptual basis of quantum physics a certain tradition has developed concerning the use of probabilistic concepts. While the necessity for these concepts is generally recognized, the tradition in question holds that the actual probabilistic concepts employed in quantum physics differ in certain essential respects from the probabilistic notions used in other areas of physics and in ordinary statistical contexts. This line of thought apparently originates with BORN himself (see JAMMER [5], p. 285). In some writings it seems that BOHR (for example, see [1], p. 4) has taken a similar point of view. And more recently the tradition has found a convincing spokesman in the work of HENRY MARGENAU (see [8] and references there).

Standing in opposition to this tradition is the elegant approach to quantum physics of VON NEUMANN [9], as developed by MACKEY [7] and others. I shall try to show that underlining this approach is a completely general setting for the use of probabilistic concepts, a setting that applies as well to quantum as to non-quantum physics, and in terms of which the quantum theory can be seen as a perfectly ordinary variety of statistical theory.

I intend to elaborate this general setting below by developing a formal account of statistical theories. I shall examine the application of this formal account to quantum physics and I shall take up certain questions regarding the interpretation of probability which naturally arise in this application. The result, I hope, will constitute a cogent rebuttal to the tradition that sees quantum probabilities as essentially different from "ordinary" ones, as well as constituting a positive and detailed suggestion for the role of probability in quantum physics. As such the present work lays a necessary background for, but does not

[1] This investigation has been supported in part by National Science Foundation Grant No. GS-2034.

otherwise contribute, towards answering the deep and interesting question concerning the apparent necessity for probabilistic concepts in quantum theory.

2 *Statistical Theories*

Most physical applications of probability have to do with the probability that the outcome of some experiment will have a special character or with the probability that, in a given situation, the range of a certain quantity will lie within specified bounds. The latter language, the language of quantities, can be used quite generally. That is, one can generally quantify an experimental situation in a straightforward way so that to speak of the experiment as having an outcome of a certain character is just to speak of the range of a special quantity as lying within specific bounds. Thus in the case of a coin tossing experiment one can, instead of speaking of the coin landing heads or tails up, speak of a certain quantity being confined (or not) to, say, the interval from zero to one. The advantage to this quantitative language is that it offers a uniform setting in which to deploy probabilistic notions. Since genuine physical theories provide a rich stock of physical quantities, this setting seems a natural one for physics.

What are quantities? It is common to think of a physical quantity as a way of associating numbers with objects at a given time. That is, one commonly treats a quantity as a real-valued function, whose domain is the pairing of some set of objects with temporal parameters. Thus in pre-relativistic physics one can consider mass (in grams) as a function that assigns to each particle at a given time a non-negative real number, the mass (in grams) of the particle at that time. One can, however, abstract from the temporal aspect somewhat by supposing that the object is, at any given moment, in a certain physical state. The value of the quantity, then, would not be determined by the temporal parameter but rather by the state of the object. Therefore, if the object is held fixed, one can treat each quantity as a real-valued function on the set of states. For statistical applications, however, a further generalization seems appropriate.

Probability and statistical methods are usually brought into play where, for reasons which may or may not have to do with some lack of information, one is not in a position to determine what the values are of the quantities with which one is concerned. The meteorologist, for example, cannot in general determine the temperature of a region, given its atmospheric state. He can, however, fix the range of temperatures and may offer probabilities for the temperature to be confined to the various intervals lying within this range. Similarly, given the physical state of a runner in a track meet, one is generally not able to say how long it will take for him to run a certain course. One can, however, place upper and lower bounds on his time and, perhaps on the basis of past records, make probabilistic predictions for the smaller temporal intervals that lie within these bounds. Such applications can be considered to treat quantities not as numerical-valued functions, but rather as set-valued functions. *That is, for a fixed object, I propose to think of a quantity as a function that associates with each state of the object a set of real numbers, the range of the quantity in that state.* In case the range of a quantity in a certain state contains just one number, we can speak of that number as "the value" of the quantity in that state. Otherwise, it is pointless to talk about the value(s) of a quantity. The meaningful focus of discourse is on a quantity's range.

This general concept of a quantity is motivated by concern for probabilistic applications. Such applications involve the ascription of probabilities to sets of numbers. The standard vehicle for this ascription is to consider probability measures defined on the Borel sets; i.e., those sets formed from the numerical intervals by taking at most countable unions, intersections and complements. In terms of this standard setting one can introduce probability in a straightforward and classical way. I shall say that a quantity Q is a *statistical quantity* with respect to P_Q just in case (a) for every state φ the range $Q(\varphi)$ of Q in φ is a Borel set and (b) P_Q is a probability measure on the family of Borel sets. I shall read $P_Q[A]$ as "the probability that Q (or the range of Q) is confined to (or is concentrated in or lies in) A." I can now develop the notion of a statistical theory.

Consider a triple $\langle \boldsymbol{S}, \boldsymbol{Q}, \boldsymbol{P} \rangle$ where $\boldsymbol{S}$ is a set of states and where for every Q in $\boldsymbol{Q}$ and every φ in $\boldsymbol{S}$, there is some P_Q^φ in $\boldsymbol{P}$ such that Q is a statistical quantity with respect to P_Q^φ. If the situation concerning the ranges of quantities over the states is to be mirrored by the probabilities, then one should certainly have that

$$P_Q^\varphi[Q(\varphi)]=1;$$

i.e., that with probability one the quantity Q in state φ is confined to its range in φ. But this necessary condition by no means captures all the information at hand. For not only is Q, in state φ, confined to its range, it is moreover not necessarily confined to any smaller set. That is, we want the probabilities to reflect the fact that, in state φ, the range $Q(\varphi)$ is the smallest set in which Q is certainly concentrated. One can approach this requirement as follows.

Let, for given state φ and quantity Q, γ_Q^φ be the family of all closed Borel sets F such that $P_Q^\varphi(F)=1$. The family γ_Q^φ is not empty since it contains, for instance, the set of all real numbers. Define F_Q^φ by

$$F_Q^\varphi=\bigcap \gamma_Q^\varphi.$$

One can readily check that γ_Q^φ has the finite intersection property and hence that F_Q^φ is non-empty. Since γ_Q^φ is a family of closed sets F_Q^φ is closed and is, therefore, a Borel set. One can now show that F_Q^φ belongs to γ_Q^φ; i.e., that

$$P_Q^\varphi[F_Q^\varphi]=1.$$

Notice, first, that for any Borel set B and set F in γ_Q^φ

$$P_Q^\varphi[B]=P_Q^\varphi[B\cap F].$$

For $P_Q^\varphi[B]=P_Q^\varphi[B\cap F']+P_Q^\varphi[B\cap F]$ (where F' is the complement of F) by the additivity of probability measures. But

$$0\leqq P_Q^\varphi[B\cap F']\leqq P_Q^\varphi[F']=0.$$

Notice, secondly, that for any Borel set B, if $(B\cap F_Q^\varphi)$ is empty, then $P_Q^\varphi[B]=0$. For in order that the antecedent hold there must be some F in γ_Q^φ such that $(B\cap F)$ is empty. But then

$$P_Q^\varphi[B]=P_Q^\varphi[B\cap F]=0.$$

The result now follows readily. Take any F in γ_Q^φ.

Since F_Q^φ is a subset of F one has that

$$1=P_Q^\varphi[F]=P_Q^\varphi[F-F_Q^\varphi]+P_Q^\varphi[F_Q^\varphi].$$

But $P_Q^\varphi[F]=1$ and since $(F-F_Q^\varphi)$ is disjoint from F_Q^φ, $P_Q^\varphi[F-F_Q^\varphi]=0$, by the second remark. Hence

$$P_Q^\varphi[F_Q^\varphi]=1.$$

Thus the set F_Q^φ is the smallest closed Borel set to which it is certain that in state φ the quantity Q is confined. It is natural to require, therefore, that $Q(\varphi)$ coincide with F_Q^φ.

I shall summarize this discussion in the following definition. A *statistical theory* is a triple $\langle \boldsymbol{S}, \boldsymbol{Q}, \boldsymbol{P}\rangle$ where $\boldsymbol{S}$ is a set of states, $\boldsymbol{Q}$ is a set of quantities and $\boldsymbol{P}$ is a set of real-valued functions subject to the following conditions:

(1) for every φ in $\boldsymbol{S}$ and Q in $\boldsymbol{Q}$ there is a P_Q^φ in $\boldsymbol{P}$ such that Q is a statistical quantity with respect to P_Q^φ, and

(2) for every φ in $\boldsymbol{S}$ and Q in $\boldsymbol{Q}$, $Q(\varphi)=F_Q^\varphi$.

I should point out that, according to (2), each quantity is well-defined on every state; i.e., the domain of each quantity is the entire set $\boldsymbol{S}$ of states. This differs, for example, from the concept of a statistical theory set out by KOCHEN and SPECKER [6] who would allow the quantities to be partial functions on the set of states. This difference merely reflects a different attitude towards quantities. KOCHEN and SPECKER take the narrow view of quantities as numerical-valued functions, whereas I take them more generally as set-valued functions.

The notion of a statistical theory is intended as a standard framework in which to cast the diverse applications of probability. It is thus a format for modeling the statistical aspects of natural phenomena. Some examples will indicate the use and scope of this format.

One can treat the whole class of coin tossing experiments, for example, by introducing as states the possible physical states of a coin, as determined by size, shape, density distribution, etc. (I assume here that each coin would be tossed by the same device in a standard way. This is an important assumption that I shall return to in Section 4.) There is only one quantity Q of interest, one that reflects the heads/tails behavior of the coin, and we can construe it formally in a fairly arbitrary way. A natural way is to restrict Q so that its range, for any state, will be one of the three non-empty subsets of $\{0, 1\}$, with 0 corresponding to tails and 1 to heads. Then we can take $\boldsymbol{P}$ to be the family of discrete probability measures on the subsets of $\{0, 1\}$ and let $Q(\varphi)$ be determined by (2). We shall thereby capture all the biased as well as unbiased coin experiments and express the probability involved, quite properly, as a function of the initial state of the coin.

In a similar way we could codify a rudimentary theory of pollution. The states might correspond to intervals of population density for a given region. The quantities could correspond to certain pollutants in the atmosphere, associating with each population density interval a range in the concentration level of a pollutant. Each probability measure would give the probable distribution of a pollutant within the expected range, as a function of population density gradients. Of course, more refined theories can be imagined.

I shall take statistical mechanics as a final example. Suppose that there is given a certain mechanical system. The set of states for statistical mechanics consists of the various probability measures φ on the family of Borel subsets of the phase space of the system. The set of quantities is determined in the following way by the set of classical dynamical variables of the system. Corresponding to each dynamical variable $\boldsymbol{q}$ there is a quantity Q defined by

$$Q(\varphi) = \{\boldsymbol{q}(x)/x \in S_0\}$$

where

$$S_0 = \bigcap \{S/S \text{ is closed and } \varphi(S) = 1\}.$$

Thus in (mixed) state φ, the range of Q is just the smallest set of numbers that $\boldsymbol{q}$ is certain to take on as values. For state φ and quantity Q, the measure P_Q^φ is determined in the obvious way from the measure φ; viz.,

$$P_Q^\varphi[A]=\varphi[\boldsymbol{q}^{-1}(A)],$$

where $\boldsymbol{q}^{-1}(A)$ consists of all the points x in phase space such that $\boldsymbol{q}(x)$ is in the set A. The resulting statistical theory provides a general version of statistical mechanics which can be suitably restricted, for example to states derivable from the Liouville measure, in physically realistic ways. (See MACKEY [7] for a development of these ideas.)

3 *Quantum Mechanics as a Statistical Theory*

Quantum mechanics readily fits into the format of statistical theories. In the von Neumann formulation a Hilbert space is associated with each quantum mechanical system. One can treat pure and mixed states in a uniform manner by taking for the states of the statistical theory the density operators on the Hilbert space. Ignoring superselection rules, the quantities correspond to the self-adjoint operators on the space in a way to be discussed below. Given a self-adjoint operator Q with spectral measure

$$S\to\varepsilon(S)$$

one can define P_Q^φ for each state φ and Borel set B by

$$P_Q^\varphi[B]=\operatorname{tr}[\varphi\,\varepsilon(B)].$$

Under this definition each P_Q^φ is a probability measure on the family of Borel sets. The operator Q can then be considered as a quantity, whose domain is the set of states, by defining $Q(\varphi)$ to be the smallest closed Borel set whose probability, under P_Q^φ, is one. This definition accords with (2) and links the states, quantities, and probability measures of quantum mechanics so as to form a statistical theory.

This way of construing quantum mechanics lumps together pure with mixed states and discrete with continuous quantities. The first of these amalgamations slurs over an important controversy concerning the physical interpretation of mixed states and their differences, if any, from the corresponding pure states. Although the resolution of this controversy has important bearings on the so-called problem of measurement (see VAN FRAASSEN [3] and FINE [2]). I do not think it affects the character of quantum mechanics as a statistical theory, and so I shall not pursue the controversy here. The unified treatment of discrete and continuous quantities, however, I consider a special virtue of this construal of quantum mechanics and I should like to suggest why this is so.

In the usual von Neumann way with quantum mechanics one is accustomed to saying that a quantity Q (= self-adjoint operator) has a value λ in state (= unit ray) φ just in case φ is an eigenstate of Q with eigenvalue λ. It then follows that Q has no value whatever in a state φ which is not an eigenstate of Q. Thus if Q has a discrete spectrum, Q takes no value in any state that is a non-degenerate superposition of its eigenstates with distinct eigenvalues. If Q has a continuous spectrum (I ignore the part discrete part continuous cases), however, then there are no eigenstates for Q at all. So a continuous quantity, it turns out, just has no values whatsoever in any state. (VON NEUMANN and others sometimes express this by saying that continuous quantities cannot be measured exactly in any state. That language, however, is quite misleading. For a glance at the demonstration shows that the concept of measurement or observation is nowhere involved. The result is, rather, a simple fact about the character of the probability measures that the theory associates with continuous quantities.) To adopt this view of quantum physics, therefore, commits one to saying that a free electron – for example – is at any time nowhere, that it has no momentum and no energy. And it does not make a difference whether the statement is made before or after a measurement for one of these quantities, since at no time does a continuous quantity have a value. Plainly, physicists do not generally speak this language; i.e., it is not customary to adopt this point of view. Rather one knows that the spread of the continuous as well as the discrete quantities is restricted in certain states and that these restrictions are precisely determined by the probabilities given by the theory. Thus the uncertainty formulae imply that if the range of position of a particle is very small, then the (linear) momentum range (in the same direction) is very large. This has the empirical consequence that when we know for sure that the particle is confined to a small interval, we expect and indeed find that momentum is spread out over a large range. That is to say, the way one normally treats the quantities of quantum physics is the way I have set out above. One treats them as set-valued, statistical quantities. In this respect the discrete and continuous quantities are quite on par.

I shall conclude this section with a discrete example that is sometimes used to suggest the paradoxical character of quantum mechanics. Far from being paradoxical, it will turn out that the interpretation of this example in the framework of statistical theories is quite straightforward and ordinary. The example concerns the quantization of angular momentum. Consider a particle in a central field, say the electron in the hydrogen atom, and consider the Z-component, L_Z, of its angular momentum. If the total angular momentum is $\sqrt{2}$ (i.e., quantum number $l=1$, in units of $\hbar/2\pi$) then there are just three eigenstates, corresponding to $L_Z=1$, 0, or -1. Suppose the particle is in the state φ corresponding to $L_Z=1$. One can expand φ in a non-degenerate superposition of the three eigenstates of $L_{Z'}$, where $Z \to Z'$ is a plane rotation of the Z-axis through some angle A, and one can compute the probability that the angular momentum is in the direction of the positive Z'-axis (i.e., $L_{Z'}=1$). Take $A=\pi/2$ in order to get some nice numbers. Then although in state φ the Z-component of the angular momentum vector points in the direction of the positive Z-axis, one has that with probability $\frac{1}{4}$ the angular momentum vector points in a perpendicular direction!

The usual response here is to say that the probabilities are not well-defined until after a measurement has been carried out. It is then suggested that measuring for $L_{Z'}$ will alter the state φ and thus avoid the supposed clash of probabilistic predictions. (See TOMONAGA [13], p. 261ff. This response also seems to be behind Sneed's attempt to treat quantum probabilities as probabilities conditional on the performance of a measurement [12].)

Let us look at this example from the point of view of statistical theories. In the notation I have been using, one has the following information:

$$\begin{array}{ll} \text{(a)} & L_Z(\varphi) = \{1\}. \\ \text{(b)} & P^{\varphi}_{L_Z}[\{1\}] = 1. \\ \text{(c)} & P^{\varphi}_{L_{Z'}}[\{1, 0, -1\}] = 1. \\ \text{(d)} & L_{Z'}(\varphi) = \{1, 0, -1\}. \\ \text{(e)} & P^{\varphi}_{L_{Z'}}[\{1\}] = \tfrac{1}{4}. \end{array}$$

According to (a) in state φ the quantity L_Z has range $\{1\}$ (i.e., we can say that in φ the value of L_Z is 1) and thus (b) confirms that with probability one L_Z is confined to $\{1\}$. According to (c), in state φ the quantity $L_{Z'}$ is confined to $\{1, 0, -1\}$ and one can readily see that this is the smallest set to which $L_{Z'}$ is certainly confined. Hence (d) says that the range of $L_{Z'}$ in state φ is $\{1, 0, -1\}$. Finally (e) asserts that in state φ the probability is $\frac{1}{4}$ that $L_{Z'}$ is concentrated in the set $\{1\}$. The paradox, presumably, is supposed to reside in the different probabilities assigned in (b) and (e) from one and the same state φ to one and the same set ("value") $\{1\}$. But as (a) and (d) demonstrate, the quantities L_Z and $L_{Z'}$ are different; they assign different sets of numbers to the same state φ. It seems to me that one would expect that the probability measures associated with different quantities would themselves be different. Thus the different probabilities assigned in (b) and (e), which may seem paradoxical from a certain point of view, turn out to be the expected reflection of a difference in quantities when viewed through the lens provided by the framework of statistical theories. One need not have recourse to the supposed disturbances brought about by measurements in order to understand why there are different probabilities here. (For a similar case HOOKER [4] has argued that measurement disturbances cannot, physically, account for the differences.)

4 *Understanding Probability*

The case of azymuthal quantization just considered illustrates a general feature of the employment of probabilities in a statistical theory. Quite generally in a state φ a certain quantity Q will have range S. Then one can partition S as $S = A \cup B$, where A and B are disjoint, and have

$$P_Q^\varphi[A \cup B] = 1$$

whereas

$$P_Q^\varphi[A] \neq 1 \quad \text{and} \quad P_Q^\varphi[B] \neq 1.$$

Thus although in state φ the quantity Q is confined to the set $(A \cup B)$ and it is neither the case that Q is confined to A nor is it the case that Q is confined to B, one is nevertheless prepared to assign probabilities to these latter eventualities. For example, in the preceding case if $A = \{-1, 0\}$ and $B = \{1\}$ one has

$$P_{Lz'}^\varphi[A \cup B] = 1$$

as in (c), but

$$P_{Lz'}^\varphi[A] = \tfrac{3}{4} \quad \text{and} \quad P_{Lz'}^\varphi[B] = \tfrac{1}{4}.$$

In such circumstances it is clear that one cannot easily give an ignorance interpretation to the probability functor; that is, I cannot say that the probability measures (in some sense) my ignorance of the true situation. For I may well know that $S = A \cup B$ is the range of Q. Thus I may be sure that Q is not confined either to A or to B. So, in speaking of the probability that Q is confined to A (or to B) I cannot plausibly be held to be expressing my ignorance of the situation. Indeed, I am already in command of all the facts that are relevant.

Not only is an ignorance interpretation of the probabilities here inappropriate, a straightforward frequency account seems out of the question as well. More precisely, one cannot interpret

$$P_Q^\varphi[A] = r \qquad (\text{for } 0 < r < 1)$$

in such a way as to yield the conclusion that in a large ensemble of systems, each of which has state φ, roughly $100\,r$ % of the systems will have Q confined to A. Such an interpretation would contradict the fact that if $A \cup B$ is the range of Q in φ and $P_Q^\varphi[B] \neq 0$, then in state φ it is not the case that Q is confined to A. In the quantum mechanical application an interpretation sanctioning the above percentages would, for example in the case of a particle "tunneling through" a potential barrier, violate well-established conservation laws.

What, then, are we to make of the probabilistic assertions? There is, I believe, a tempting metaphor that is sometimes used here to ease the difficulty. Suppose we think of

$$P_Q^\varphi[A \cup B] = 1$$

as saying that Q is spread throughout its range, $Q(\varphi)=A\cup B$. Then there is no reason to hold that Q must be spread uniformly over $A\cup B$. It may, indeed, be more concentrated in some regions than in others. The degree of concentration is precisely what the probability measures. Thus I can sensibly say that although Q is not entirely confined to either one of A or B, it is—for example—three times more concentrated in A than it is in B. This would yield

$$P_Q^\varphi[A]=\tfrac{3}{4} \quad \text{and} \quad P_Q^\varphi[B]=\tfrac{1}{4}.$$

The underlying picture here is of an extended spatial entity (like a lake or a cloud) that occupies the region $A\cup B$ with more of it in A than in B. The virtue of this picture is that it makes the ascription of probability signify the presence of a real property that the object already has in state φ; namely the variable concentration of the "Q-cloud" over $A\cup B$. The defect of this picture is that it is not readily applicable to those abstract quantities, like momentum, energy, or spin, that do not lend themselves to this spatial metaphor. It is especially difficult to see how discrete quantities fit into this picture. What is it to say, for example, that the direction of the angular momentum vector is "spread out" over $\{-1, 0, 1\}$, being "more concentrated" in $\{-1, 0\}$ than in $\{1\}$?

I think, therefore, that one must look for a different way to understand the ascriptions of probability. This way may be susceptible to a spatial metaphor in some cases, for instance where probabilities are determined by the concentrations in a diffraction pattern or by the intensities of spectral lines. But our understanding of probability should not be made to depend on such a metaphor.

In the framework of statistical theories, for a fixed object and quantity, the associated probability measure is given as a function of the state of the object. Thus it is by virtue of being in a certain state that the various probabilities arise. It seems natural, then, to think of the probabilities as reflecting certain features of the state of the object. Thus, a certain feature of state φ is reflected in the fact that

$$P_Q^\varphi[A]=r.$$

We might call this "the Q in A with probability r" feature of φ.

What more can be said about this feature? Surely the following is correct. If $r=1$, then the only possibility for the range of Q is a subset of A. (In the discrete case, one can say that the only possible values for Q are among the numbers in A.) If $r=0$, then it is not possible that Q lies in A. If $0<r<1$, then—first of all—the state φ is a state of chance for Q to lie in A. That is, if $0<r<1$, then there is a chance both that Q lies in A and that Q does not lie in A. Moreover, this state of chance is one that admits probabilities. That is, the possible pattern of distributions of Q (in or out of A) is sufficiently regular so that it can be assessed by means of a probability measure, P_Q^φ. Finally, the specific assessment

$$P_Q^\varphi[A]=r$$

directs us to the specific distribution patterns that could possibly occur for an object in state φ. These are the patterns in which, roughly, $100\,r\,\%$ of the pattern corresponds to the range of Q lying in A. (I realize that this last statement is vague in the way frequency accounts of probability are notoriously liable to be. But it is not my intention here to deal with the difficulties of frequency accounts. I seek only to suggest that in the context of statistical theories—and hence in the context of quantum theory—the theoretical ascriptions of probability may lead one to expect patterns of results distributed with certain regularities.)

This way of trying to spell out the content of probabilistic assertions has made repeated reference to possibilities. My construal of probability here is essentially modal in the following sense. To say that

$$P_Q^\varphi[A]=r$$

does not tell us what the situation is in state φ with regard to the "event" that Q is confined to A. It tells us rather about what the possibilities are (as judged from φ, one might say) for this event. In so doing I suppose that I might leave myself open to the charge of explaining the difficult by means of the obscure. For we sometimes tend to feel that talk of possibilities is necessarily elusive. But if the reader will go over the last two sentences, both of which involves modalities, he may perhaps be persuaded that *possibility* need not be elusive. I believe that the possibilities involved in understanding probabilistic assertions are not at all obscure, at least no more so than the usual probabilistic discourse from which they derive. I shall try to illustrate why this is so.

Suppose I hold in my hand or place on the table a coin for which I employ the sort of statistical theory outlined in Section 2. On the basis of the physical state of the coin I may be prepared to say that the probability for its landing heads up is $\frac{1}{2}$. Suppose I have a spinner twirling around on a board, portions of which are either red or green. On the basis of the physics of the situation, I may be inclined to say that the probability for the spinner resting on "red" is 1. Suppose I prepare an electron in a state φ so that $L_Z=1$. On the basis of the usual quantum mechanical uses of the state function, I am prepared to say that $1, 0, -1$ are the only possible values for $L_{Z'}$. Moreover, I am prepared to say that the probability for the angular momentum vector to be oriented in a way that corresponds to $L_{Z'}=1$ is $\frac{1}{4}$.

Each of these cases involves an object in a certain state and the probability, in that state, for the range of a certain quantity to be confined in a specified way. In each case I may assume that when the probabilistic assertion is being made, although the object is in the given state, the quantity involved is not confined to the specified range. (Yet the probability for its being so confined, as in the case of the twirling spinner, may even be 1.) In each case my suggestion is that the probabilistic assertion is a way of projecting from the given state to possible patterns for the range of the quantity to be confined (or not) as specified. For the coin this possibility depends on tossing the coin in a standard way to land on a standard surface. For the twirling spinner this possibility depends merely on allowing sufficient time to elapse for the spinner to come to rest. For the electron the possibility depends on arranging an interaction between the electron and some other system for which there is a quantity that correlates with angular momentum, in such a way that one can infer the probability measure for $L_{Z'}$ from the distribution of this correlated quantity. (This is an unwieldy description of what I take a quantum measurement to be.) In each case the possibility merely involves the projected outcomes of a tacitly understood test procedure. In each case the projection is based on the responses that are expected to occur when objects having the physical features described by the state are subjected to the environment of a test, as determined by the appropriate physical theories.

I believe that this says a great deal that is sensible about "the Q in A with probability r" feature of a state φ. In particular, the difficulty with which we began, the conjunction of $A \cup B = Q(\varphi)$ with

$$P_Q^{\varphi}[A] = \tfrac{1}{4}, \quad P_Q^{\varphi}[B] = \tfrac{3}{4}$$

is readily understood. The test procedure if initiated on objects in state φ is bound to reflect Q confined to $A \cup B$ and it is not bound to reflect Q confined to any smaller subset. But roughly one-quarter of the results of this test procedure will reflect Q confined to B. (I use this cautious language in order to leave room for tests that may involve only very indirect correlations with the object.) It may happen, of course, that particular objects in state φ do in fact have Q confined to some measurably smaller subset of $A \cup B$. For instance, my coin may already be lying heads up on the surface. Whereas my twirling spinner cannot be at rest, nor can the angular momentum of my electron be oriented both in the direction of the positive Z and Z' axes. Thus whether Q does in fact lie in some smaller subset of its range will depend on the particular Q, φ and physical theory involved. This situation further reflects the modality inherent in a statistical theory. It cannot be expected that the state description will yield all the relevant information about the object. At best it will yield just those items of information that can be connected in a lawlike way. (The view of probability offered here can be considered a version of the "propensity interpretation" associated with POPPER [10]. It is also similar to, but a little more cautious—and looser—than, the explicitly modal account of VAN FRAASSEN in [3]. For criticisms of this way with probability see SKLAR [11].)

In terms of these suggestions for understanding probability in a statistical theory, I think one can see the source of the spatial metaphor—that Q is spread throughout its range but concentrated more in some areas than in others. The metaphor sees that ascriptions of probability relate to objective features of the object. In my view these are precisely the features given by its state description. The metaphor then misses the modal features of probability and seeks to attribute to the object in its antecedent state a smearing out of results that, in my view, are merely projected as patterns of test responses. The metaphor thus fuses antecedent with consequent conditions in a vague picture that is, I hope, brought into sharper focus by a physical analysis along the lines of the approach suggested above.

5 *Conclusion*

I have attempted to show that there is a general account of statistical theories inherent in the approach of VON NEUMANN and MACKEY to the foundations of quantum mechanics. This account is profitably elaborated by employing the notion of a quantity as a set-valued function. This account makes use only of the classical set-theoretic formulation of probability theory. And this account of statistical theories is broad enough to encompass the most diverse scientific uses of probability, including both statistical mechanics and quantum mechanics. I have not tried here to resolve general questions about the problem of interpreting probabilistic assertions. But I have tried to indicate that there is a modal account of probabilities that fits naturally into the scheme of statistical theories. According to this account statements of probability function to restrict the possible patterns that may emerge from an environment that is generated by certain standard tests. I have tried to argue that the modality here is not problematic but is merely the sort of projection from initial conditions that physical sciences have always employed.

It is my hope that the framework of statistical theories provides a background in terms of which one can gain some insight into the variety of puzzles that surround the use of probability both in quantal and in non-quantal science. In particular, I would hope that this framework offers a coherent setting for investigating the (apparent) necessity for recourse to probabilities in the quantum domain.

References

1. BOHR, NIELS: Essays 1958/1962 on atomic physics and human knowledge. New York: Interscience Publishers 1963.
2. FINE, ARTHUR: Insolubility of the quantum measurement problem. Phys. Rev. D, vol. II, **12**, pp. 2783–2787 (1970).
3. FRAASSEN, BAS VAN: Measurement in quantum mechanics as a consistency problem. (Preprint.)
4. HOOKER, CLIFFORD: The nature of quantum mechanical reality: Einstein versus Bohr, to appear in R.G. Colodny (ed.), Paradigms and paradoxes: The challenge of the quantum domain. Englewood Cliffs, N.J.: Prentice-Hall, Inc. (forthcoming).
5. JAMMER, MAX: The conceptual development of quantum mechanics. New York: McGraw-Hill Book Co. 1966.
6. KOCHEN, SIMON, SPECKER, E. P.: The problem of hidden variables in quantum mechanics. J. Math. and Mech. **17**, 58–87 (1967).
7. MACKEY, GEORGE W.: Mathematical foundations of quantum mechanics. New York: W.A. Benjamin, Inc. 1963.
8. MARGENAU, HENRY, COHEN, LEON: Probabilities in quantum mechanics, in: M. Bunge (ed.), Quantum theory and reality, pp. 71–89. Berlin-Heidelberg-New York: Springer 1967.
9. NEUMANN, JOHN VON: Mathematical foundations of quantum mechanics. Princeton, New Jersey: Princeton University Press 1955 (trans. R.T. Beyer).
10. POPPER, KARL R.: The propensity interpretation of probability. British J. Philos. Sci. **10**, 25–42 (1959).
11. SKLAR, LAWRENCE: Is probability a dispositional property. J. Philos. **67**, 355–366 (1970).
12. SNEED, JOSEPH: Quantum mechanics and classical probability theory. Synthese **21**, 34–64 (1970).
13. TOMONAGA, SIN-ITIRO: Quantum mechanics, vol. II. Amsterdam: North-Holland Pub. Co. 1966.

7 The Localization Problem

ANDRÉS J. KÁLNAY [1]

Departamento de Física, Facultad de Ciencias,
Universidad Central de Venezuela, Caracas (Venezuela)

1 Introduction

The localization problem in relativistic quantum mechanics consists in finding (i) the operator representative X^k of position and/or its eigenstates (called localized states), (ii) their properties and (iii) the representatives and properties of variables related to position such, as time, proper time and velocity. We devote this review mainly to position and briefly to variables like velocity. Time and proper time will be discussed only when relevant to position.

Much work was done on position and velocity in relativistic quantum mechanics: see Refs. [1–87]. The search started when BREIT [1] noticed that the eigenvalues of the velocity deduced from DIRAC's multiplicative position operator for the electron were equal to $\pm c$ (velocity of light), a strange property whose discovery was followed by the one of the Zitterbewegung by SCHRÖDINGER [2]. In spite of the many papers published on the localization problem there is no proposal of a solution on which general agreement can be reached.

Instead of reviewing the different papers we shall describe lines of research. Individual contributions will be only summarized when useful to illustrate a given approach or its evolution in time. We shall not be able to review individual articles which are far removed from the main approaches. The views of the author on the ways of tackling the problem will appear in Section 6.

We shall only consider the free particle case, with which most of the literature is concerned, except in a few cases.

2 Conventions

Greek letters take on the values 0, 1, 2 and 3 of spacetime indices; latin letters take on the values 1, 2 and 3. We denote $E_p = +(\boldsymbol{p}^2 + m_0^2)^{\frac{1}{2}}$

[1] *Present address:* Sección Física, IVIC. Apartado 1827, Caracas, Venezuela.

where m_0 stands for the rest mass. We call s the proper time. The notations and conventions not explicitely stated are those used by MESSIAH [88].

The 3-position operator will be called $\boldsymbol{X}$ and the 4-position X^μ but x_D^μ will be used to denote DIRAC's position. The coordinates of a frame of reference will be labelled x^μ or a^μ.

As there is no sure knowledge *a priori* on the value of $[X^i, X^j]$ we should distinguish [3] *1-localized states* (eigenstates of only one component X^k),

$$X^k \varphi_{a^k} = a^k \varphi_{a^k} \tag{2.1 a}$$

from *3-localized states*

$$\boldsymbol{X} \varphi_{\boldsymbol{a}\lambda} = \boldsymbol{a} \varphi_{\boldsymbol{a}\lambda} \tag{2.1 b}$$

which only exists if for all i, j we have $[X^i, X^j] = 0$. Here λ stands for the additional degeneration (if any) and will be omitted further on.

We use the natural system of units ($c = \hbar = 1$), except in some self-evident places.

3 *The Formal Properties Approach* [6, 7, 9–12, 14–19, 21]

3.1 The Properties. We list now a set of properties that seem quite reasonable for position. Different subsets of this set were assumed explicitely or implicitely by different authors. Depending upon the author, a special subset was used: (i) as a set of axioms such that the position operator could be deduced from them, or (ii) as a set of properties which a previously assumed position operator had and which was used to show that the assumed operator was a good candidate, or (iii) as a test which, on failure, showed that a candidate for a position operator should be disregarded. We call *the formal properties approach* the one which uses (i), (ii) or (iii) as the basic tool to solve the localization problem [89].

The properties will be grouped in the following way [90]:

a) Set of transformation properties under homogeneous transformations of coordinates:

- *a1)* The position $\boldsymbol{X}$ is a 3-vector under space rotations. Almost all position candidates satisfy this requirement. See however Refs. [4, 5].
- *a2)* Relativistic generalizations of (a1). They can have different forms:
  - *a2.1)* An operator X^0 exists such that X is a 4-vector [6, 7, 9, 10, 91].
  - *a2.2)* The existence of X^0 is not assumed but still a relativistic covariant meaning is given to the 3-vector position [12].

a3) The position $\boldsymbol{X}$ is a polar vector. This is a general requirement [92].

a4) The position $\boldsymbol{X}$ is invariant under time reflection [14–16].

a5) Transformation properties of the localized states (see Section 7).

b) Set of transformation properties under translations.

b1) As linear momentum is the generator of translations it is imposed [12, 14–19] $[X^i, P^j]=i\delta_{ij} I$.

b2) Relativistic variants of (b 1) [20, 93].

b3) Transformation properties of the localized states (see Section 7).

c) Set of properties of $\boldsymbol{X}$ *in relation with the angular momentum*

c1) Let us call $\boldsymbol{M}$ the total angular momentum. It is assumed that an operator $\boldsymbol{S}$ exists such that

$$\boldsymbol{L}=\boldsymbol{X}\times\boldsymbol{p}, \qquad \boldsymbol{M}=\boldsymbol{L}+\boldsymbol{S} \tag{3.1}$$

and such that the components of $\boldsymbol{S}$ have the same commutation relations that the total angular momentum have.

c2) Eq. (3.1) is assumed but other properties are looked on $\boldsymbol{S}$, such that $\boldsymbol{S}$ be a conserved quantity for the free particle (in constrast with DIRAC's spin) [6].

c3) The observables $\boldsymbol{S}$ and $\boldsymbol{X}$ are taken to be simultaneously measurable [19], i.e. $[\boldsymbol{S}^i, \boldsymbol{X}^j]=0$.

c4) Relativistic generalizations of (c 1), (c 2) and/or (c 3) [6, 7, 10, 11].

d) Set of properties characterizing velocity, acceleration, etc.

d1) The relation $d\boldsymbol{X}/dt=\boldsymbol{p}E_p^{-1}$ (or variants of it, is assumed [14–16, 19].

d2) Several facts support $(1-\boldsymbol{v}^2)^{\frac{1}{2}}_{\text{op}}=\gamma^0$ so that instead of (d 1) one sometimes tries to obtain [6, 7]

$$d\boldsymbol{X}/dt=(1-\boldsymbol{v}^2)^{\frac{1}{2}}\,\boldsymbol{p}/m_0=\gamma^0\,\boldsymbol{p}/m_0 .$$

d3) The relation $(dX^\mu/ds)\,\psi=(p^\mu/m_0)\,\psi$ for the 4-velocity is expected to hold [7, 9].

d4) The condition [18] $d^2\boldsymbol{X}/dt^2=0$ is imposed.

d5) One also tries to satisfy the relation $(d^2 X^\mu/ds^2)\,\psi=0$ for the 4-acceleration [7, 9].

d6) Generalizations of the requirements (d 1) to (d 5) are used when interactions are present [6, 7].

e) Set of properties about the compatibility of components of position.

e1) It is required [14, 15, 18] that $[\boldsymbol{X}^i, \boldsymbol{X}^j]=0$ for all i, j.

e2) The stronger requirement $[X^\mu, X^\nu]=0$ for all μ, ν is also imposed [10].

f) Set of properties about the hermiticity.

f1) The hermiticity of the 3-position has almost always been assumed. See however Sections 8 and 9 and references quoted there.

f2) Variables related to $\boldsymbol{X}$ (as X^0) are taken to be Hermitian. See however Refs. [6–8, 22].

g) Range property.

g1) It is required that the particle (or antiparticle) position operator has its domain and range in the corresponding state space. This is usually expressed by means of projection operators on particle and antiparticle state vector space [14–16, 23, 24].

3.2 Discussion. Most ot the properties mentioned above seem to provide a set of natural requirements on position. However, it is certain that *there is at least one subset which is false.* This can be seen from the fact that different authors chose different subsets as axioms from which the position operator and/or the localized states could be deduced, but the results contradict each other. *If standard quantum mechanics is assumed to hold, no solution exists such that it agrees with the whole of reasonable requirements.* For definiteness, let us only consider two examples: JORDAN and MUKUNDA [12] showed that for the (negative charge) electron the most general solution which agrees with (a 1), (a 2.2), (b 1) and (f1) fails to satisfy the property (e 1). If electrons and positrons are simultaneously considered the most general solution of (a 1), (a 2.2), (b 1), (e 1) and (f1) exists but does not agrees with (d 1).

4 *The Center of Mass Approach* [10, 23, 25–30]

The idea is the following [23, 25, 26]: Let us have a classical relativistic continuum mass density $\rho(x)$ and its center of mass

$$\boldsymbol{X}_{\mathrm{CM}}=\int \rho(\boldsymbol{x})\, \boldsymbol{x}\, d^3 \boldsymbol{x} / \int \rho(\boldsymbol{x})\, d^3 \boldsymbol{x} \tag{4.1a}$$

and express it as a function of other classical observables $A_1^{\mathrm{cl}}, \ldots, A_n^{\mathrm{cl}}$,

$$\boldsymbol{X}_{\mathrm{CM}}=\boldsymbol{f}(A_1^{\mathrm{cl}}, \ldots, A_n^{\mathrm{cl}}). \tag{4.1b}$$

Let us have now a quantum relativistic particle whose position operator (or "the nearest approach to ... position" [23]) be $\boldsymbol{X}$; if its measurable values are called $\boldsymbol{x}$ and if $\rho(\boldsymbol{x})$ is its density in a given state, then the corresponding mean value is

$$\langle X \rangle = \int \rho(x)\, x\, d^3 x / \int \rho(x)\, d^3 x. \tag{4.2a}$$

Eq. (4.2a) looks *formally* like (4.1a). This suggest that instead of Eq. (4.1b) it may be right to use the relation

$$\langle X \rangle = f(A_1, \ldots, A_n) \tag{4.2b}$$

in the quantum case, so that the position operator would be

$$X = f(A_1, \ldots, A_n) \tag{4.3}$$

where A_k is the quantum variable whose classical partner is A_k^{cl}.

This procedure would seem to give a way of finding the quantum partner of the classical position (but the ordering of non-commutative operators A_k in Eq. (4.3) must be considered). However, the analogy that supports this method does not necessarily hold: The functions f used in Eqs. (4.1b) and (4.3) may differ by terms that go to zero in the classical limit. However the main difficulty of the method lies in the fact that the classical relativistic definition of center of mass is ambiguous [23]. That this is the case can be seen from the fact that several densities, such as the energy density ($c=1$) and the proper mass density, become the same non relativistic mass density when the low velocities limit is taken. Several definitions of the classical relativistic center of mass can be given and in each case they correspond to different function f in Eq. (4.1b), which, in turn, leads to different possibilities for the quantum formula (4.3) [23]. Thus, the center of mass approach, though atractive, does not deliver a definite solution of the quantum localization problem.

5 *The Unitary Transformations Approach* [31–39]

Let us start with the standard representation

$$i\gamma^\mu\, \partial\psi(x, t)/\partial x^\mu = m_0\, \psi(x, t) \tag{5.1}$$

of DIRAC's equation for the electron. At the begining of the localization problem DIRAC considered the multiplicative operator x_D^k as the representative of position, a hypothesis that seems now disregarded. However, if the three components of the "true" position X^k exist and commute with each other, then instead of affirming that the multiplicative operator is not a position, it can be said that the standard representation is not suitable for identifying the multiplicative operator with the position representative. In other words, it can be said that another representation,

$$\psi' = U\psi, \qquad U^+ U = U U^+ = I \tag{5.2}$$

exists such that the multiplicative operator X'^k in this new representation is the representative of position. Then, transforming back the X'^k to the standard representation, the position representative in this last representation is obtained.

Foldy, Wouthuysen and Tani [31] considered the transformation for the electron (free particle case)

$$U=(m_0 I+\gamma\cdot p+E_p)[2E_p(E_p+m_0)]^{-\frac{1}{2}} \tag{5.3}$$

which carries the Hamiltonian to the form

$$H'=\gamma^0 E_p, \tag{5.4}$$

This representation is very convenient because in it the positive and negative energy solutions can be uncoupled most easily, and also because the nonrelativistic limit can be better studied (this is specially true when electromagnetic interactions are introduced). Foldy and Wouthuysen [31] chose this particular U as the one to be used in order to find the position by the method mentioned above. The solution they found is attractive because of its properties. It agrees with (a1), (a3), (a4), (a5), (b1), (b3), (c1), (c2), (c3), (d1), (d4), (e1), (f1) and (g1) (cf. Section 3) and it coincides (or better, almost coincides [27]) with the one obtained by others using different approaches (e.g. the one by Newton and Wigner [42]: see Section 7). However it seems not to be the definitive solution because it violates the equivalence of inertial frames (see Section 7). Nevertheless, since the solution appears to be "almost right" it proved attractive to find the position by the same method but with a change of the operator U. This, together with the extension of the original Foldy-Wouthuysen procedure to other particles, originated *the unitary transformations approach.* A remarkable synthesis of different generalizations was obtained by Hanus *et al.* [34]. However, with the increasing number of solutions it became clear that this method, in spite of its interest, is not the right tool to sort out the right from the wrong solutions of the localization problem.

6 A Stop-Over

In Sections 3, 4 and 5 we reviewed briefly several of the approaches used to solve the localization problem. Before continuing in the next sections with other lines of approach, it is worthwhile to discuss the following question. From the previous sections we saw (i) that formal approaches based on analogies do not provide us with a definite solution to the problem and (ii) that to follow the physicist's intuition is very risky in this case, because properties that seem natural for the position to satisfy are in conflict with each other, so that at least some of them are wrong (cf. Section 3).

If experimental information concerning sharp localization were available, we could use it to select the right solution. But no such precise data are available, partly because theoreticians have failed to develop the theory of the localization to a point where the properties of position that would enable us to measure it would show up [3]. Under such circunstances, it seems to us that two lines of research should be stressed:

A) To see whether we know a physical property such that (i) it is supported by experimental evidence independently of hypotheses about microscopic localization and (ii) an (at least partial) answer to the localization problem could be deduced from it [3].

B) To throw light on implicit assumptions that were used in past attempts to solve the problem. To call in doubt any implicit or explicit hypothesis no matter how respectable it may seem [13]. Perhaps one of such hypotheses (such as the full validity of relativistic quantum mechanics) is wrong and causes the present embarrassment about localization. Of course, this open minded point of view must include the possibility that the very concept of localization has no sense in microphysics – but then the problem of localization in macrophysics would arise [3]. Also it may be that the space-time has a discret structure (see the work of Bopp and others [40]). The generalized localizability in the sense of Jauch and Piron [41] must also be considered.

Attempts related to A and B will be discussed respectively in Sections 7 and 9.

7 *The Equivalent Frames of Reference Approach* [42–59]

7.1 Newton and Wigner's Paper [42]. We regard the classical paper by Newton and Wigner (NW) as the most fundamental one on the localization problem, even though we find discrepancies between our work and their final conclusions. The importance of the NW paper lies in the fact that *the equivalent frames of reference approach* has a sound form. We believe that this approach is the best of the class A (Section 6) that we have up to now.

NW assumes for a particle the following [94]:

NW1: The set S_0 of states localized at the origin of space-time is a linear set.

NW2: This set is invariant under (a) spatial rotations about the origin O and (b) under space and time reflections.

NW3: States localized at $(t=0, \boldsymbol{x})$ and $(t=0, \boldsymbol{x}')$ with $\boldsymbol{x} \neq \boldsymbol{x}'$ are orthogonal.

NW4: A condition of mathematical good behaviour.

From these assumptions NW deduce a unique position operator $\boldsymbol{X}_{\mathrm{NW}}$ and the localized states, and this is done for any spin. For spin zero, $\boldsymbol{X}_{\mathrm{NW}}$ takes the form in the p-representation, Heisenberg picture, $t=0$,

$$\boldsymbol{X}_{\mathrm{NW}}=i(\partial/\partial \boldsymbol{p})-i\,\boldsymbol{p}/(2E_p^2). \tag{7.1}$$

The 3-localized states in $(t=0, \boldsymbol{a})$, i.e. the eigenfunctions in

$$\boldsymbol{X}_{\mathrm{NW}}\,\varphi_{\boldsymbol{a}}^{\mathrm{NW}}=\boldsymbol{a}\,\varphi_{\boldsymbol{a}}^{\mathrm{NW}} \tag{7.2}$$

are, in this case,

$$\varphi_{\boldsymbol{a}}^{\mathrm{NW}}=(2\pi)^{-\frac{3}{2}}\,e^{-i\boldsymbol{a}\cdot\boldsymbol{p}}\,E_p^{\frac{1}{2}}. \tag{7.3}$$

Axiom NW1 is necessary in order that Eq. (7.2) could be written. Axiom NW3 expresses that the probability of finding the particle at $t=0$ at the point $\boldsymbol{x}$ is zero if it is certainly at $\boldsymbol{x}'\neq\boldsymbol{x}$ at the same time; it is not necessary if the particle has extension [60]. For an analysis of Axiom NW4 and ambiguities related to it see Ref. [44].

In our opinion the most remarkable is Axiom NW2, being basic for the development of *the equivalent frames of reference approach*. We shall discuss its physical content in the next subsection.

7.2 Physical Discussion of Axiom NW 2. The physical meaning that, in our view, is implicit in NW's second axiom will be discussed here. Consider a particle localized at the origin O of space time with respect to a given inertial frame of reference Σ [95]. Call Σ' a second frame related to Σ by a spatial rotation around O. Coordinates change when going from Σ to Σ' but this is not the case of O: its coordinates $(0,0,0,0)$ are invariants of the transformation. With regard to Σ the particle is certainly at $x^\mu=0$. Then it follows that for its coordinates x'^μ in Σ' it is also certain that $x'^\mu=0$. Otherwise the descriptions of the same state from Σ and Σ' would be mutually inconsistent, and this is impossible because spatially rotated frames are physically equivalent. Now we use the well known fact that it is the same to leave fixed the physical system and to rotate the frame of reference as to leave fixed the frame and to rotate the physical system in the opposite direction. It follows that if ψ belongs to the set S_0 of states localized at the origin of Σ, then a spatial rotation performed on the physical system (while the frame remains unchanged) transforms ψ into ψ' which also belongs to S_0. But this is just saying that the set S_0 is invariant under rotations. This is part (a) of Axiom NW2. Part (b) can be similarly discussed *if* frames related by reflections are physically equivalent.

As axiom NW2, part (a), gives a rotationally invariant meaning to localization, it can be called the *axiom of rotational invariance of localization*. It follows from this discussion that *this invariance is a selfconsistency requirement of localization* which follows from the physical equivalence of spatially rotated frames of reference.

7.3 A Necessary Generalization: Lorentz Invariance of Localization. In a relativistic theory not only spatially rotated frames are physically equivalent, but also those related by a homogeneous Lorentz transformation. Then it follows from a reasoning similar to the previous one that *localization should also be Lorentz invariant*, i.e. that S_0 be invariant under homogeneous Lorentz transformations.

The approach that uses the physical equivalence of frames of reference as a tool to solve the localization problem belongs to the class A mentioned in Section 6, because the physical equivalence of inertial frames has an independent and sound experimental verification. It gives a sure basis to tackle the localization problem.

7.4 Why Can the NW Solution Not be the Definitive One? Consequences. The NW solution cannot be the right one because, as recognized in the original paper and shown several times by other authors, their solution violates the Lorentz invariance of localization. This cannot happen in a theory that deals (as NW do) with relativiste systems. As the Lorentz invariance of localization has a sure basis, i.e. the physical equivalence of inertial frames, we are forced to give up at least one of the explicit or implicit axioms that support NW's theory, no matter how reasonable it may look. If localization has sense in microphysics, we have among others the following possibilities:

a) To drop orthogonality (Axiom NW 3). This rejection can have a physical basis (cf. Section 7.1).

b) Related to (a), not to require hermiticity of $\boldsymbol{X}$. This *can* be done without conflict with quantum mechanics (see Section 8).

c) To change the mathematical condition NW 4. See GALINDO [44].

d) Not to require the commutativity of components of position. NW's axioms are only possible if

$$[X^i_{\mathrm{NW}}, X^j_{\mathrm{NW}}]=0, \qquad \forall i,j, \tag{7.4}$$

but this is an assumption without sure support [3, 48, 96] because (i) it has not been tested experimentally at small distances and (ii) there exist well known physical variables, such as the angular momentum, whose components do not commute at the quantum level. In any case, what could be expected from the classical limit of commutators and from the fact that *ordinary* classical particles have zero Poisson brackets $\{x^i, x^j\}=0$, is that *perhaps*

$$\lim_{\hbar\to 0} \hbar^{-1}[X^i, X^j]=0, \qquad \forall i,j \ (?) \tag{7.5}$$

which is a weaker requirement. The question mark stands for the fact that we cannot exclude the possibility that, in the classical limit, quantum relativistic particles exhibit strange properties.

e) To reformulate localization in such a way that even if $[X^i, X^j]=0$ the sentence "the set of states localized at the origin of space time" has no more sense (see Section 9.1).

f) To change the present form of relativistic quantum mechanics [97] (cf. Section 9.2).

In the remainder of Section 7 we shall work under the hypothesis that relativistic quantum mechanics is right and without reformulating localization, i.e. excluding possibilities (e) and (f).

7.5 Consequences of the Lorentz Invariance of Localization. PHILIPS [49, 51, 52, 98] postulated the Lorentz invariance of localization in a form suitable only if the position components are compatible with each other. He found that if a mathematical conjecture he assumed was correct, then for the spin zero case the states localized at $\boldsymbol{a}$ would be what are now called *Philips' class I localized states*

$$\varphi_{\boldsymbol{a}}^{\mathrm{Phil}}=(2\pi)^{-\frac{3}{2}}e^{-i\boldsymbol{a}\cdot\boldsymbol{p}}. \tag{7.6}$$

He also concluded that no position operator exists because he had no hermitian solution. However non-hermitian operators are acceptable (cf. Section 8) and using this fact it was shown in Ref. [50] that this operator exists and it was constructed. In Ref. [54] the mathematical conjecture assumed by PHILIPS was shown to be wrong so that the localization problem appeared again to have many solutions.

However this lack of uniqueness was solved in Ref. [3] where it was shown that the Lorentz invariance of localization (in its less restricted form, neither assuming nor rejecting compatibility of the componentes of position) was sufficient (without any further hypothesis than quantum mechanics) to provide a unique solution at least for the electron and the spinless mesons. (The uniquenes was shown up to a parameter G for the electron.) The reason why here, with weaker requirements, a definite solution was obtained, is that it was realized that Eq. (2.1a) – which follows from quantum mechanics if position exists – provides a legitimate restriction on the transformation properties of the localized states because $\boldsymbol{X}$ can only be constructed with the operators of the theory and they have definite transformation properties under the Lorentz group.

The research done in Ref. [3] was extended to the zero mass case [13] and the method was positively tested in the non relativistic limit [48, 55] (for other work in the non relativistic limit see Refs. [99, 100]). It was proved [56] that the parameter G that remained unknown for the electron had the value zero and that as a consequence, and at least for the electron, the neutrino, the photon and the case $m_0 = \text{spin} = 0$, the position operator was unique in each case and that it could be put into the same form (p representation, Heisenberg picture, $t=0$)

$$\boldsymbol{X} = \Lambda(i\,\partial/\partial\boldsymbol{p})\,\Lambda \tag{7.7}$$

in all cases, where Λ is the corresponding projection operator into the allowed state vector space. For spin zero, the localized states are Philips class I localized states. For the other cases they were found but its explicit form is complicated [3, 13].

The position operator (7.7) is non-hermitian (and even non-normal [101]) which does not contradict quantum mechanics (Section 8), but its eigenvalues are real [3, 13, 102]. It agrees with the formal properties (Section 3) (a1), (a3), (a4), (a5), (b1), (b3), (c2), (c4), (d1), (d4) and (g1). (Thus, the right set of formal properties follows from the equivalence of inertial frames.) Property (e1) is only satisfied for spin zero. For the properties (d3), (d5) (with $\mu = 1$, 2 and 3) and (d6) no answer can be given without further elaboration. When analyzed as a binary variable (Section 8) its first part is equal for spinless masive mesons to NW's operator [42] and for the electron [3, 56] to that of BUNGE's [6]. The second part is a constant of the motion. The remarkable properties of NW's operator (essentially equal to FOLDY-WOUTHUYSEN's [27, 31, 42]) were given in Section 5. BUNGE's operator is

$$X^\mu = x^\mu_D + (i\,\gamma^\mu/2\,m_0), \tag{7.8}$$

and reconciles formal covariance with classical analogues [6, 7]. Its properties are (a1), (a2.1), (a3), (a4), (b1), (b3), (c2), (c4), (d2), (d3), (d5), (d6) and (f1). When electromagnetic interactions are introduced, it has reasonable properties [6, 7].

Once the explicit form of the localized states is known [3, 13] all physical properties of localization can, in principle, be also known. *An urgent task is to use this to deduce an experimental procedure to measure position.*

7.6 Four Vectors and Covariance. The work done for constructing possible 4-vector position operators was expected to give a Lorentz invariant localization. In spite of the great interest of the work done in this direction, the 4-vector approach is not a safe one within the frame of standard quantum mechanics. Following a preliminary discussion in Ref. [3] this will be shown in Ref. [56]. Let us mention here only one difficulty: in order to get Lorentz invariant localization from 4-position it is necessary that X^0 be the time [3], but it is well known that in standard quantum mechanics (including the relativistic case) the time is a c-number [104] (see e.g. Refs. [26, 103, 105]) while the space componentes X^k are operators which is inconsistent with the 4-vector operator properties of X^μ.

FLEMING [61] has nicely clarified the relation between the description of quantum processes relative to different inertial frames, but his transformation of position 3-vectors into 4-vectors does not solve the lack of Lorentz invariance of e.g. NW's solution. (See footnote 3 of Ref. [3]). See also Refs. [62–64].

Let us conclude by mentioning that the 3-position (7.7), which is consistent with the Lorentz-invariance of localization, is *not* the space part of a 4-vector [3, 13].

7.7 Other Invariance Approaches. BECKMANN [65] showed how to find Lorentz invariant localized states for systems composed of a definite number (greater than one) of particles, but his method cannot be extended to single particle states.

There are indications that the study of the imbedding of space-time in higher dimensions is worthwhile. CASTELL [43, 57], KASTRUP, FLINT and WILLIAMSON and MADUEMEZIA [58] looked for the invariant solution of the localization problem in this context.

8 A Note on Hermiticity

Possible position operators were rejected in the past because they were not Hermitian (see e.g. Refs. [107, 108]). This is risky because non Hermitian operators

$$A = A_I + i A_{II}, \qquad A_I = A_I^+, \qquad A_{II} = A_{II}^+ \tag{8.1}$$

can represent legitimate quantum variables [109, 110]. The reason can be easily seen for normal operators [101]. Their eigenvalues are complex numbers. The usual argument that complex numbers are not measurable is incorrect: complex numbers $a = a_I + i\, a_{II}$ are ordered pairs (a_I, a_{II}) of real numbers and the result of an experiment can be such a pair. Non normal operators also have physical meaning as binary variables (A_I, A_{II}) but the discussion is more intricate [110].

Given an unknown physical variable, to assume that it is a standard hermitian variable instead of a normal but non hermitian, or even a binary non normal one, is a hypothesis that should not be assumed as correct without due justification. This is especially true for such a baffling case as that of the quantum relativistic position, whose nature we *a priori* ignore.

The spin zero position rejected in Ref. [107] because it is not Hermitian is *identical* with the one given in Eq. (7.7) (for spin zero $\Lambda = I$) which is the only one consistent with Lorentz invariance.

9 *Questionable Assumptions*

This section is devoted to some approaches to B (Section 6) i.e. to the possibility that some of the implicit assumptions used here to fore to solve the localization problem might be wrong.

9.1 Why Should Position be Point-Like. This is the possibility (e) mentioned in Section 7.4. In Ref. [60] a reformulation of the notion of localization such that the values of the position be regions of diameter not lower that $1/m_0$ was proposed. If this is correct for the so called elementary particles, then to ask for their point position would be a question without sense and perhaps this is the cause of the embarrassment in the localization problem. The physical meaning of the reformulated position may be some intrinsic property of the particles or the fact that the "elementary" particles are not really elementary. These ideas were developed to a preliminary extent in Refs. [60, 50, 66, 67, 111]. They were used by RECAMI and BALDO to solve a controversy on tachyon physics [68] and applied by OLKHOVSKY and RECAMI [69] to the space-time descritption of collision processes.

Non-hermitian operators are suitable in this approach [60] because to each coordinate a pair of variables are associated: the location of a center of the region and a diameter; this pair operates naturally as a binary variable (Section 8).

9.2 Quantum Mechanics. In this Subsection some attemps related to the possibility (f) of Section 7.4 will be shortly reviewed. JOHNSON [64] and BROYLES [70, 71] created two modified forms of relativistic quantum mechanics which, though having remarkable differences, have some fundamental ideas in common [113]. Usually one starts from standard relativistic quantum mechanics and try to find place in it for position. Unfortunately, as we have seen, we cannot do that without losing some properties we would like to keep, such as orthogonality or the existence of a 4-position. JOHNSON and BROYLES give instead priority to the existence of a hermitian 4-vector position as well as of linear and angular momentum and relate the variables in such a way that the properties one would like are preserved; then they reformulate relativistic quantum mechanics so that it does not contradict these assumptions. They give up the c-number property of time and change the Hilbert space to be used. Their localized states are not eigenstates of the mass,

$$[m_{op}^2, X^\mu] \neq 0, \qquad m_{op}^2 = p_\nu p^\nu. \tag{9.1}$$

Instead, all masses, including imaginary ones, contribute to a localized state. (We remark the coincidence with a result of BARUT and MALIN [72] who work within the frame of standard relativistic quantum mechanics: they found that a strongly localized state of a Majorana particle includes more than one mass.) JOHNSON's kets evolve in proper time, where a c-number proper time is used, while BROYLES considers a view of the whole space-time at once and looks for probability distributions of *events*.

We caution against Eq. (9.1) which is essential for both papers because, although their formalism gives automatically a Lorentz invariant description of localization, the position operator given in Eq. (7.7) which follows from the Lorentz invariance of localization, is such that

$$[m_{op}^2, X_{(7.7)}^\mu] = 0. \tag{9.2}$$

Nevertheless we consider that the work by JOHNSON and BROYLES, as well as that of ROSENBAUM [74] (who generalizes quantum mechanics to a super-Hilbert space with a similar purpose) is a deep approach to the task of solving the localization problem. We should have an open mind to the possibility that the solution will come from a change in some of the features of relativistic quantum mechanics, regardless of how accustomed to them we are. We mention here again an interesting paper by CASTELL [57]: his use of higher dimensions to solve localization leads to a generalization of quantum mechanics.

The above mentioned changes proposed for the present form of quantum mechanics are related to the difficulty in reconciling the c-number character of time with a manifestly covariant description of localization. In this context a new scheme of particle quantization stated by HANUS [75] has special interest, as well as the work done on FLEMING ideas [61–63, 114].

Acknowledgment. The author is indebted to Prof. P. L. TORRES for a critical reading of the manuscript.

References and Footnotes

1. BREIT, G.: Proc. Nat. Acad. Sci. U.S.A. **14**, 553 (1928).
2. SCHRÖDINGER, E.: Preuss. Akad. Wiss. Berlin **24**, 418 (1930). – GUTH, E.: Ann. Physik **20**, 309 (1962). – HÖNL, H., PAPAPETROU, A.: Z. Physik **116**, 153 (1940).
3. KÁLNAY, A. J.: Phys. Rev. **1** D, 1092 (1970).
4. COESTER, F.: Phys. Rev. **129**, 2816 (1963).
5. FRONSDAL, C.: Phys. Rev. **113**, 1367 (1959).

6. Bunge, M.: Nuovo Cimento **1**, 977 (1955). See also Refs. [7], [8], [22] and [84].
7. – Kálnay, A.J.: Progr. Theoret. Phys. (Kyoto) **42**, 1445 (1969). – Bunge, M.: Ciencia e Investigación **14**, 311 (1958). – Plahte, E.: Progr. Theoret. Phys. (Kyoto), Suppl. **4**, 246 (1966); **4**, 291 (1966). – Yamasaki, H.: Progr. Theoret. Phys. (Kyoto) **31**, 324 (1964); **39**, 372 (1968).
8. – British J. Philos. Sci. **9**, 39 (1958).
9. Kálnay, A.J., MacCotrina, E.: Progr. Theoret. Phys. (Kyoto) **42**, 1422 (1969); Erratum and Addenda **43**, 569 (1970).
10. Bacry, H.: Phys. Letters **5**, 37 (1963); – J. Math. Phys. **5**, 109 (1964).
11. Lugarini, G., Pauri, M.: Nuovo Cimento **57**A, 299 (1967).
12. Jordan, T.F., Mukunda, N.: Phys. Rev. **132**, 1842 (1963).
13. Kálnay, A.J.: Phys. Rev. **3**D, 2357 (1971).
14. Mathews, P.M., Sankaranarayanan, A.: Progr. Theoret. Phys. (Kyoto) **26**, 499 (1961).
15. – – Progr. Theoret. Phys. (Koyto) **27**, 1063 (1962). – Sankaranarayanan, A., Good, R.J., Jr.: Phys. Rev. **140**, B509 (1965).
16. Garrido, L.M., Sesma, J.: Amer. J. Phys. **30**, 887 (1962).
17. Langbein, W.: Z. Physik **198**, 304 (1967).
18. Bardakci, K., Acharya, R.: Nuovo Cimento **21**, 802 (1961).
19. Berg, R.A.: J. Math. Phys. **6**, 34 (1965). – Candlin, D.J.: Nuovo Cimento **37**, 1396 (1965).
20. Finkelstein, R.J.: Phys. Rev. **75**, 1079 (1949).
21. Heitler, W.: The quantum theory of radiation. Oxford 1954, p. 106.
22. Kolsrud, M.: Phys. Norveg. **2**, 149 (1967).
23. Pryce, M.H.L.: Proc. Roy. Soc. Ser. A **195**, 62 (1948).
24. Mathews, P.M., Sankaranarayanan, A.: Progr. Theoret. Phys. (Kyoto) **26**, 1 (1961). See also Mathews, P.M., Sankaranarayanan, A.: Progr. Theoret. Phys. (Kyoto) **32**, 159 (1964).
25. Papapetrou, A.: Prakt. Akad. Athénes **14**, 540 (1939); **15**, 404 (1940).
26. Møller, C.: Communications of the Dublin Institute for Advanced Studies A, No. 5 (1949).
27. Chakrabarti, A.: J. Math. Phys. **4**, 1223 (1963).
28. Shirokov, Iu. M.: Soviet Physics JETP **6**, 664 (1958).
29. Shirokov, Iu. M.: Z. Èksper. Teoret. Fiz. **21**, 748 (1951).
30. Melvin, M.A.: Rev. Mod. Phys. **32**, 477 (1960).
31. Foldy, L.L., Wouthuysen, S.A.: Phys. Rev. **78**, 29 (1950). – Tani, S.: Soryuschiron Kenkyu **1**, 15 (1949); – Progr. Theoret. Phys. (Kyoto) **6**, 267 (1951).
32. Bose, S.K., Gamba, A., Sudarshan, E.C.G.: Phys. Rev. **113**, 1661 (1959).
33. Hanus, W., Janyszek, H., Rakowski, A.: Bull. Acad. Polon. Sci. Sér. Sci. Math. Astronom. Phys. **16**, 253 (1968).
34. – – – Acta Phys. Polonica **34**, 1071 (1968).
35. Baktavatsalou, M.: Nuovo Cimento **25**, 964 (1962); – Compt. Rend. Acad. Sci. Paris **258**, 474 (1964).
36. – Cahiers Physique **155**, 1 (1963).
37. Pac, P.Y.: Progr. Theoret. Phys. (Kyoto) **22**, 857 (1959). See also Ref. [24].
38. Janyszek, H., Rakowski, A.: Acta Phys. Polonica **35**, 529 (1969).
39. Case, K.M.: Phys. Rev. **95**, 1323 (1954). – Kálnay, A.J.: Thesis, Chap. 3, Suppl. to vol. **11**, Notas de Fisica, Centro Brasileiro de Pesquisas Fisica, Rio de Janeiro (1964). – Azcarraga, J.A. de, Boya, L.J.: J. Math. Phys. **9**, 1689 (1968). – Azcarraga, J.A. de, Oliver, L.: J. Math. Phys. **10**, 1869 (1969). – Case, K.M.: Phys. Rev. **95**, 1323 (1954). – Chakrabarti, A.: Nuovo Cimento **18**, 617 (1960). – Sesma, J.: J. Math. Phys. **7**, 1300 (1966). – Strocchi, F.: Nuovo Cimento **33**, 1437 (1964). – Parravicini, G., Sparzani, A.: Nuovo Cimento A**66**, 579 (1970). See also Beckers, J.: Phys.

Rev. **3** D, 952 (1971), FESHBACH, H., VILLARS, F.: Rev. Modern Phys. **30**, 24 (1958) and the review paper by DE VRIES, E.: Fortschr. Physik **18**, 149 (1970).

40. BOPP, F.: S.-B. Bay. Akad. Wiss., Math.-Naturw. Kl., H. 2, 12 (1967); – Z. Physik **200**, 117 (1967); **200**, 142 (1967); **200**, 133 (1967). – AHMAVAARA, Y.: J. Math. Phys. **6**, 87 (1965); **6**, 220 (1965). – HELLUND, E. J., TANAKA, K.: Phys. Rev. **94**, 192 (1954). – ROJANSKY, V.: Phys. Rev. **97**, 507 (1955).
41. JAUCH, J. M., PIRON, C.: Helv. Phys. Acta **40**, 559 (1967).
42. NEWTON, T. D., WIGNER, E. P.: Rev. Modern Phys. **21**, 400 (1949).
43. CASTELL, L.: Nuovo Cimento **46** A, 1 (1966).
44. GALINDO, A.: Nuovo Cimento **37**, 413 (1965).
45. WIGHTMAN, A. S.: Rev. Mod. Phys. **34**, 845 (1962).
46. LUNN, M.: J. Phys. A, Ser. II, **2**, 17 (1969). – WIGHTMANN, A. S., SCHWEBER, S. S.: Phys. Rev. **98**, 812 (1955). – FONDA, L.: Nuovo Cimento A **56**, 1095 (1968).
47. WEIDLICH, W., MITRA, A. K.: Nuovo Cimento **30**, 385 (1963).
48. KÁLNAY, A. J., TORRES, P. L.: Phys. Rev. **3** D, 2977 (1971). See also References quoted in this paper.
49. PHILIPS, T. O.: Phys. Rev. **136**, B 893 (1964); – Thesis, Princeton University, 1963, available through Microfilm-Xerography-University Microfilms, Inc., Ann. Arbor, Mich.
50. GALLARDO, J. A., KÁLNAY, A. J., STEC, B. S., TOLEDO, B. P.: Nuovo Cimento **49**, 393 (1967).
51. SCHRÖDER, U.: Ann. Physik **14**, 91 (1964).
52. KNIGHT, J. M.: J. Math. Phys. **2**, 459 (1961). – LICHT, A. L.: J. Math. Phys. **4**, 1443 (1963).
53. PHILIPS, T. O., WIGNER, E. P.: In: Group theory and its application, edit. by E. M. Loebl. New York: Academic Press Inc. 1968.
54. GALLARDO, J. C., KÁLNAY, A. J., RISEMBERG, S. H.: Phys. Rev. **158**, 1484 (1967).
55. KÁLNAY, A. J., TORRES, P. L.: Internat. J. Theoret. Phys. (L) **3**, 167 (1970).
56. – – (Work in progress.)
57. CASTELL, L.: Nuovo Cimento **49**, 285 (1967).
58. KASTRUP, H. A.: Phys. Rev. **143**, 1021 (1966). – FLINT, H. T., MARJORIE WILLIAMSON, E.: Proc. Roy. Soc. (London), Ser. A **207**, 308 (1951). – MADUEMEZIA, A.: Nuovo Cimento 51 A, 585 (1967).
59. SHIROKOV, M. I.: Ann. Physik **10**, 60 (1962).
60. KÁLNAY, A. J., TOLEDO, B. P.: Nuovo Cimento **48**, 997 (1967).
61. FLEMING, G. N.: Phys. Rev. **137**, B 188 (1965); **139**, B 963 (1965).
62. – J. Math. Phys. **7**, 1959 (1966).
63. MCDONALD, S. C.: J. Math. Phys. **11**, 1158 (1970).
64. JOHNSON, J. E.: Phys. Rev. **181**, 1755 (1969); – Phys. Rev. **3** D, 1735 (1971).
65. BECKMANN, P.: Nuovo Cimento **27**, 868 (1963). – CLOSE, F. E., OSBORN, H.: Phys. Rev. **2** D, 2127 (1970). – OSBORN, H.: Phys. Rev. **176**, 1514 (1968).
66. GALLARDO, J. A., KÁLNAY, A. J., STEC, B. A., TOLEDO, B. P.: Nuovo Cimento **48**, 1008 (1967).
67. KÁLNAY, A. J.: Bol. Imaf (Universidad Nacional de Córdoba, Argentina) **2**, 41 (1966).
68. BALDO, M., RECAMI, E.: Lett. Nuovo Cimento **2**, 643 (1969).
69. OLKHOVSKY, V. S., RECAMI, E.: Lett. Nuovo Cimento **4**, 1165 (1970). – RECAMI, E.: Rend. Accad. Naz. Lincei, Cl. Sci. Fis., Mat. Natur., Ser. 8, **49**, 77 (1970).
70. BROYLES, A. A.: Phys. Rev. **1** D, 979 (1970).
71. – Space-time position operators and spin 1/2 particles (Preprint), University of Florida.
72. BARUT, A. O., MALIN, S.: Rev. Modern Phys. **40**, 632 (1968).
73. BOPP, F.: Z. Angew. Phys. **1**, 387 (1949).
74. ROSENBAUM, D. M.: J. Math. Phys. **10**, 1127 (1969).

75. HANUS, W.: Reports on Math. Physics **1**, 245 (1971).
76. ACHARYA, R., SUDARSHAN, E.C.G.: J. Math. Phys. **1**, 532 (1960).
77. AMREIN, W.O.: Helv. Phys. Acta **42**, 149 (1969).
78. BECK, G.: Rev. Fac. Ci. Univ. Coimbra **10**, 66 (1942). – SZAMOSI, G.: Nuovo Cimento **20**, 1090 (1961).
79. CORBEN, H.C.: Phys. Rev. **121**, 1833 (1961).
80. FRADKIN, D.M., GOOD, R.JR.: Rev. Modern Phys. **33**, 343 (1961).
81. GIAMBIAGI, J.J.: Nuovo Cimento **16**, 202 (1960). – NISHIYAMA, T.: Progr. Theoret. Phys. (Kyoto) **7**, 417 (1952).
82. HANUS, W.: Acta Phys. Polon. **29**, 463 (1966).
83. KOBA, Z.: Progr. Theoret. Phys. (Kyoto), Suppl. **8**, 1 (1958); Nuovo Cimento **3**, 1 (1956); – **3**, 214 (1956).
84. KOLSRUD, M.: Phys. Math. Univ. Oslo, Report No. 27 (1965); – Nuovo Cimento **39**, 504 (1965).
85. SCHLIEDER, S.: Comm. Math. Phys. **1**, 265 (1965).
86. We have no place to discuss two important subjects related to the localization problem: Local observables and localized fields. See e.g. ARAKI, H.: J. Math. Phys. **5**, 1 (1964); – Progr. Theoret. Phys. (Kyoto) **32**, 844 (1964). – JAFFE, A.M.: Phys. Rev. **158**, 1454 (1967). – KHORUZHII, S.S.: Theoret. and Math. Physics USSR **1**, 73 (1969). – WIGHTMAN, A.S.: Ann. Inst. H. Poincaré, Sect. A **1**, 403 (1964).
87. BLOKHINTSEV, D.I.: Report International Centre for Theoretical Physics, Trieste, IC/67/36; – Acta Phys. Acad. Sci. Hungar. **22**, 307 (1967). – GALINDO, A.: Helv. Phys. Acta **41**, 257 (1968). – GARRIDO, L.M., PASCUAL, P.: Nuovo Cimento **12**, 181 (1959). – GÜRSEY, F.: Report Çekmece Nuclear Research and Training Center, Istambul, "Localization properties of the unitary representations of the Poincaré group in field theory". – HEITLER, W.: Proc. Roy. Irish Acad., Sect. A **49**, 1 (1943). – KORFF, D.: J. Math. Phys. **5**, 869 (1964). – LICHT, A.L.: J. Math. Phys. **7**, 1656 (1966). See also references quoted in this paper. – MARX, E.: Nuovo Cimento **57** B, 43 (1968); **60** A, 669 (1969); **60** A, 683 (1969). – PRYCE, M.H.L.: Proc. Roy. Soc. (London), Ser. A **150**, 166 (1935). – RAZAVY, M.: Nuovo Cimento **63** B, 271 (1969). – SHIROKOV, M.I.: J. Nuclear Physics USSR **2**, 332 (1965). – STEINMANN, O.: Comm. Math. Phys. **7**, 112 (1968). – WAKANO, M.: Progr. Theoret. Phys. (Kyoto) **35**, 1117 (1966).
88. MESSIAH, A.: Mécanique Quantique. Paris: Dunod Cie. 1960.
89. If an author chose (i) as his approach and confirms his results through (ii), he will be only quoted when writing the properties he used as axioms.
90. The set is not exhaustive. Some properties were not listed because of difficulties of systematization.
91. This requirement has different forms: If X^0 is an operator, it means that X be a 4-vector operator; if X^0 is a c-number, it only means the appropriate transformation properties of mean values.
92. However it should not be imposed on the neutrino. See Ref. [13].
93. However if (b 1) is rejected the behaviour of position under translations is not clear.
94. For simplicity we shall not discuss the notion of elementary systems, for which NW's axioms are imposed. See Ref. [42].
95. In NW's paper, for simplicity, it is not taken into account the fact that eigenstates of a variable with continuum spectra (as position) cannot be strictly observed, but Wightman's paper (Ref. [45]) implies that under essentially the same ideas of NW's paper but taking into account this point in a more rigurous way, position operator exists and is unique, which coincides with the results of NW.
96. JAUCH, J.M.: Foundations of quantum mechanics. Addison-Wesley Publishing Company 1968.
97. We are indebted to Prof. M. BUNGE for calling our attention to this possibility (M. BUNGE, private communication).

98. Starting from second quantization Schröder found related results (cf. Ref. [51]). Knight (Ref. [52]), working in quantum field theory obtained non orthogonal but Lorentz invariant localized states such that they may not contain a finite number of particles.
99. Inönü, E., Wigner, E. P.: Nuovo Cimento **9**, 705 (1952).
100. Hamermesh, M.: Ann. Physics **9**, 518 (1960).
101. Normal operators are those whose Hermitian and pure anti-hermitian parts commute.
102. Notice that in relativistic quantum mechanics the state vector scalar product is not the usual one (see e.g. Ref. [42]) so that the standard theorem "reality of eigenvalues implies Hermiticity" does not hold (cf. Ref. [3]).
103. Pauli, W., in: Handbuch der Physik, Bd. 5/1, edit. by S. Flügge. Berlin-Göttingen-Heidelberg: Springer 1958.
104. The physical reason is that the time in which a quantum experiment is done is measured as the classical time of the (macroscopic) frame of reference.
105. Schrödinger, E.: Berl. Ber. p. 238 (1931).
106. An exceptional case is the one of Bunge's X^μ (which formally is a 4-vector) whose time component is not a c-number. The exception arises from the fact that in Bunge's theory the description of localization is not done with X^μ alone but with the pair (X^μ, x_D^μ) where x_D^0 is the ordinary c-number time (cf. Refs. [6] and [7]). This implies that the above mentioned objection do not apply to Bunge's X^μ.
107. Schweber, S. S.: An introduction to relativistic quantum field theory, chap. 3. New York 1964.
108. Peres, A.: Lett. Nuovo Cimento **1**, 837 (1969).
109. Kemble, E. C.: The fundamental principles of quantum mechanics, p. 249. New York 1958.
110. See Section 3 of Ref. [60].
111. For a correction of a mistake in Ref. [60] see the third footnote on p. 1434 of Ref. [9].
112. For a related work on linear momentum see Gruber, G. R.: Found. Physics **1**, 227 (1971).
113. For earlier work related to these, see Stuckelberg, E. C. G.: Helv. Phys. Acta **14**, 322 (1941); **15**, 23 (1942).
114. Fleming, G. N.: Phys. Rev. **1** D, 542 (1970).

8 Semantic Foundation of the Quantization Program [1]

ARTHUR KOMAR

Department of Physics, Belfer Graduate School of Science, Yeshiva University, New York, N.Y.

1 Introduction

It is a well-known result of VON NEUMANN [1] that the representation of the commutation relations of the basic dynamical variables of quantum theory is unique modulo a unitary transformation. This result, valid only for systems having finite numbers of degrees of freedom, seems to imply that, for such mechanical systems, the procedure of quantization is not sensitive to the particular choice of classical canonical variables which one initially employs. This implication is in fact false. The choice of canonical variables which one employs to initiate a quantization of a given classical theory is quite critical.

The purpose of this paper is twofold: first we shall indicate the basis for the inadequacy of VON NEUMANN'S result, and then we shall elucidate the further considerations which have proved necessary to establish a successful quantum theory. Our principal interest in this latter regard stems from our concern to establish a quantum theory of gravitation. The criteria deduced for the selection of classical phase space variables preferred for quantization prove to be curiously inapplicable to the treatment of the gravitational field, which fact may account for the notable lack of success to date in the development of a quantum theory of gravitation.

2 Transformations of Operators

The dynamical variables of classical mechanics play two distinct roles. On the most primitive level, they are regarded as attributes of the physical system under investigation. Thus one frequently employs the symbols q and p to denote the position and momentum respectively, of a particular degree of freedom of a physical system. In this sense the

[1] Work supported in part by the U.S. Air Force under Grant No. AF-AFOSR 68-1524A.

dynamical variables q and p are attributes of the physical system contributing to the specifications of its instantaneous state, their evolution in time being determined by the dynamics of the physical system. The dynamical laws of classical physics, however, are found to have a striking symmetry which enables one to map physically distinct system into each other provided they have the same number of degrees of freedom. This symmetry can be characterized by an abstract group, the canonical group, namely the group of transformations which leaves invariant the symplectic form when treated as a tensor field over an even-dimensional manifold dubbed the "classical phase space". The elements of this canonical group in the infinitesimal neighborhood of the identity can be realized by arbitrary functions of the dynamical variables. The commutators of the infinitesimal elements is then realized by the Poisson brackets of the corresponding functions. In this sense the dynamical variables play a more formal role. They may be regarded as operators which generate infinitesimal mappings of classical systems upon each other. While the operator properties of the dynamical variables are more remote from our primitive physical intuition, they play a vital role in enabling us to identify at least some of the abstract symbols occurring in a given physical theory with measurements performed in nature. It should be evident that once we have permitted the possibility of different physical theories to be mapped upon each other, the identification of symbols with measurements becomes a non-trivial matter.

In quantum theory, the distinction between the operator and the attribute aspect of the dynamical variables is more transparent. The symmetry of the dynamical laws in quantum theory is that of the unitary mappings in the Hilbert space. The dynamical variables are most directly identified as operators which generate the infinitesimal unitary mappings. The commutators of the infinitesimal unitary mappings are now realized by the algebraic commutators of the abstract symbols which represent the dynamical variables. The correspondence principle, which enables us to relate at least some of the dynamical variables of a classical theory with those of the related quantum theory, employs in a vital way the intimate relationship between the operator properties of the dynamical variables and the means by which they are measured in nature. Measurements, of course, are neither "classical" nor "quantum", for those words refer only to theories by means of which we hope to understand the significance of the actual operations occurring in performing the measurements. Unlike the corresponding situation in classical theories, attributes of quantum systems cannot be realized by the same symbols which realize the operators. Rather, the attributes, that is the numerical consequences resulting from the performance of measurements, are now realized by the eigenvalues of the associated operators.

In order to initiate the procedure of canonical quantization one selects a set of fundamental classical dynamical variables q_i, p_j whose Poisson brackets are

$$\begin{aligned}[q_i, q_j] &= [p_i, p_j] = 0 \\ [q_i, p_j] &= \delta_{ij}.\end{aligned} \tag{2.1}$$

These relations are preserved under arbitrary canonical transformations, thus the formal identification of these variables is by no means unique. One then associates with each variable q_i, p_j a corresponding Hermitian operator Q_i, P_j such that their algebraic commutators satisfy

$$\begin{aligned}[Q_i, Q_j] &= [P_i, P_j] = 0 \\ [Q_i, P_j] &= i\delta_{ij}.\end{aligned} \tag{2.2}$$

(We are assuming units such that $\hbar = 1$. The factor i occurs due to our employment of Hermitian operators rather than infinitesimal unitary ones.)

Von Neumann's theorem [1] asserts that the set of operators which satisfy Eq. (2.2) is unique up to a unitary transformation. In view of the fact that the classical variables which satisfy Eq. (2.1) are unique up to a canonical transformation, it is clear that the initial selection of a canonically transformed set of classical dynamical variables would merely result in a unitarily transformed set of corresponding quantum dynamical variables. It would therefore appear that the canonical quantization of a given classical theory is insensitive to the particular choice of initial fundamental variables employed.

However, more is demanded in the procedure of quantization than the mere correspondence of basic dynamical variables. A physical system is frequently characterized by the dynamics which it satisfies. By "dynamics" we mean that there exists a well-defined relationship between the motion of a system in space and its evolution in time. Thus the dynamics of a physical system is determined by a constraining relationship between the operator which generates the mapping of the system in time, usually called the Hamiltonian, and the operators which generate the mappings of the system in space. More specifically, the dynamics of a physical system is generally characterized by a functional relationship

$$h = h(q_i, p_j) \tag{2.3}$$

where h is the Hamiltonian of the system. Thus if $h = p^2/2m$ we call such a system a free particle of mass m, while if $h = (p^2/2m) + \frac{1}{2} k q^2$ we call such a system a simple harmonic oscillator of mass m and spring constant k. It is important to note at this point that such a characterization of physical systems is not invariant under canonical transformations.

A critical step in the construction of a quantum theory is the requirement that the operator generating the evolution of the quantum system in time, also called the Hamiltonian, satisfies a constraint of the same functional form (as nearly as possible consistent with it being a Hermitian operator) as function of the fundamental quantum dynamical variables, as its corresponding classical counterpart. Thus

$$H = H(Q_i, P_i) = h(Q_i, P_j) \tag{2.4}$$

where by the second equality we mean the same functional form as occurs in the corresponding classical expression, Eq. (2.3). Thus, the Hamiltonian for a quantum free particle of mass m is $H = P^2/2m$, while the quantum simple harmonic oscillator Hamiltonian is $H = (P/2m) + \frac{1}{2} k Q^2$.

If the correspondence of dynamics as realized by Eq. (2.4), were preserved under the action of the canonical transformations of the classical variables q_i, p_j and the corresponding unitary image of these transformations acting on the quantum variables Q_i, P_j, we could assert that the procedure of canonical quantization is invariant under canonical transformations. However this is not the case. The difficulty stems from the fact that under canonical transformations the classical Hamiltonian transforms as a scalar field over the phase space spanned by the dynamical variables q_i, p_j, whereas under unitary transformations the quantum Hamiltonian transforms as an operator, that is, as a Hermitian tensor, in the Hilbert space. Thus, although the transformation law for the q_i, p_j can be coordinated with those of the Q_i, P_j, the coordination cannot be preserved for h and H. In fact, while the functional form of H is preserved under unitary transformations, the functional form of h will in general change under canonical transformations.

We shall demonstrate this critical point by considering an infinitesimal canonical transformation which is simply an extended point transformation. Let us take for our example the infinitesimal generator

$$c = \varepsilon p q^2. \tag{2.5}$$

Thus, the infinitesimal mapping generated by c is

$$\begin{aligned} \bar{q} - q \equiv \delta q = [q, c] = \varepsilon q^2 \\ \bar{p} - p \equiv \delta p = [p, c] = 2\varepsilon p q. \end{aligned} \tag{2.6}$$

Suppose that the classical system which we are considering has the Hamiltonian

$$h = \frac{p^2}{2m} + V(q). \tag{2.7}$$

We are interested in the change in the functional form of the Hamiltonian as a function of the dynamical variables due to the transformation generated by c. In view of the fact that the Hamiltonian is regarded as a scalar under canonical transformations, we find

$$h(q, p)=\bar{h}(\bar{q}, \bar{p})=\bar{h}(q+\delta q, p+\delta p). \tag{2.8}$$

We therefore find for the change in the functional form of the Hamiltonian, to the first order in ε,

$$\begin{aligned}\bar{\delta} h &\equiv \bar{h}(q, p)-h(q, p)\\ &=-\varepsilon\left(\frac{\partial h}{\partial q}\,\delta q+\frac{\partial h}{\partial p}\,\delta p\right)\\ &=-[a, c].\end{aligned} \tag{2.9}$$

For the particular choice of c, Eq. (2.5), we have

$$\bar{\delta} h=\varepsilon\left(\frac{2p^2 q}{m}-V'(q)\, q^2\right). \tag{2.10}$$

The functional form of the Hamiltonian has evidently been altered in a substantial way. This property of canonical transformations, of altering the functional form of the Hamiltonian is, of course, one of its principal and well-known virtues.

The Hermitian operator which we can associate with the classical generator of Eq. (2.5) is

$$C=\varepsilon\,\tfrac{1}{2}(PQ^2+Q^2 P). \tag{2.11}$$

The infinitesimal mapping generated by C is, via Eq. (2.2)

$$\begin{aligned}\bar{Q}-Q &\equiv \delta Q=-i[Q, C]=\varepsilon Q^2\\ \bar{P}-P &\equiv \delta P=-i[P, C]=-\varepsilon(QP+PQ).\end{aligned} \tag{2.12}$$

By comparison with Eq. (2.6) we see that thus far we have indeed performed the infinitesimal unitary transformation related to the canonical transformation generated by Eq. (2.5).

Unlike the analogous situation in classical theory, the Hamiltonian of quantum theory transforms as a Hermitian tensor, that is, it transforms via the same law as do the operators Q and P. Thus

$$\begin{aligned}\bar{H}-H &\equiv \delta H=-i[H, C]\\ &=-\varepsilon\left(\frac{P^2 Q+2PQP+QP^2}{2m}-V'(Q)\, Q^2\right)\end{aligned} \tag{2.13}$$

or, to first order in ε

$$\bar{H}=\frac{P^2}{2m}+V(Q)-\varepsilon\left(\frac{P^2 Q+2PQP+QP^2}{2m}-V'(Q)\, Q^2\right). \tag{2.14}$$

This expression corresponds to the classical change in value of the Hamiltonian, rather than the change in functional form. To obtain the change in functional form we must reexpress $\bar{H}$ in terms of the transformed operators $\bar{Q}$ and $\bar{P}$. Thus from Eq. (2.12), to order ε we have

$$\begin{aligned} Q &= \bar{Q} - \varepsilon \bar{Q}^2 \\ P &= \bar{P} + \varepsilon(\bar{Q}\bar{P} + \bar{P}\bar{Q}) \end{aligned} \tag{2.15}$$

and consequently

$$\begin{aligned} P^2 &= [\bar{P} + \varepsilon(\bar{Q}\bar{P} + \bar{P}\bar{Q})] \cdot [\bar{P} + \varepsilon(\bar{Q}\bar{P} + \bar{P}\bar{Q})] \\ &= \bar{P}^2 + \varepsilon(\bar{P}^2\bar{Q} + 2\bar{P}\bar{Q}\bar{P} + \bar{Q}\bar{P}^2). \end{aligned} \tag{2.16}$$

Substituting Eqs. (2.15) and (2.16) into Eq. (2.14) we find to first order in ε

$$\bar{H} = \frac{\bar{P}^2}{2m} + V(\bar{Q}). \tag{2.17}$$

Therefore, as we have already anticipated, there is no change in the functional form of the quantum Hamiltonian! It is this discrepancy in the behavior of the classical and quantum Hamiltonians under corresponding transformations, which gives rise to the need for assigning preferred status to some particular set of classical dynamical variables when seeking to construct the corresponding quantum theory. Only under very special transformations do the processes of canonical transformation and canonical quantization commute [2].

3 *Semantics of Dynamical Variables*

It is of interest to inquire into the method of selection of the preferred dynamical variables referred to in the previous section. The clue to their selection is the well-known dictum that one must quantize only in "rectangular coordinates". Of course, it is not a satisfactory state of affairs to have so basic a criterion stated in a coordinate-dependent manner. In fact, it becomes particularly illuminating to determine the invariant criterion for the selection of the preferred canonical variables.

The space-time which is the arena of physics has a preferred isometry group, the Galilean group for non-relativistic physics or the Poincaré group for special relativistic physics. The given isometry group singles out a preferred set of vector fields on the space which are a solution of an invariant differential equation known as Killing's equation [3]. By selecting four commuting Killing directions we obtain a preferred set of space-time coordinates determined up to the transformations available in the isometry group. The three spatial coordinates so obtained induce a preferred coordinatization of the classical configuration space, and the preferred time-like coordinate yields a corresponding coordinatization of the tangent bundle over the configuration space. Finally, if we are given a dynamics, say in the form of a Lagrangian $L(q_i, \dot{q}_i)$, we can construct the momentum p_i canonically conjugate to the chosen set of q_i via the usual relation

$$p_i \equiv \frac{\partial L}{\partial q_i}, \tag{3.1}$$

thereby obtaining a preferred coordinatization of the cotangent bundle, that is, the classical phase space. The essential point of this construction is that for the purposes of quantization a vital role is played by the assumed isometries of the space-time.

Examining the classical dynamical variables so obtained, that is, the q_i and p_j, we discover that upon viewing them in their operator aspect they generate the canonical mappings associated with the space-time isometries. In fact, upon closer inspection, we discover that it precisely this association of the preferred dynamical variables with the space-time isometries which plays a critical role in enabling us to relate their values to the output of given measuring devices [4]. Until such a relationship is established between the abstract symbols of the theory and measuring devices, our symbols are purely formal, satisfying well-defined syntactic laws of manipulation, but having no semantic content. Thus, the semantic content of a physical theory fundamentally is attained only upon establishing a relationship between the operator aspect of a given symbol and an isometry of space-time.

It now becomes clear why the program of canonical quantization places such critical emphasis on the association of certain classical commutators with corresponding quantum commutators. Not all classical commutation relations are preserved in the transition to quantum theory. An inspection of those commutation relations which are preserved intact in this transition reveals that it is precisely those commutation relations associated with the space-time isometry group! From our present perspective we therefore see that it is precisely the semantic content of the fundamental dynamical variables which is preserved by the procedure of canonical quantization. This is as it should be, for the relationship between the abstract symbols and the measurements which give them semantic content must be assumed ab initio, that is, before we set about to discover the correct (syntactic) laws which these symbols must satisfy.

Having thus identified the deeper significance of the program of canonical quantization let us complete the description of the procedure. The sub-group of the canonical group whose Lie algebra is preserved under quantization is the sub-group of linear transformations of the preferred phase space coordinates constructed above [2]. (It is isomorphic to the symplectic group and contains as a sub-group the space-time isometries.) The quantization is completed by requiring that this symplectic group be represented by unitary mappings of a Hilbert space in such a fashion that the specific algebraic form of the mapping between the tangent bundle over the configuration space and the cotangent bundle, when expressed as a function of the preferred phase space coordinates, be preserved.

4 General Relativity

In an effort to construct a quantum theory of gravitation it would appear necessary first to expose the semantic content of the basic dynamical variables occurring in the general theory of relativity. In that theory, the preferred space-time group is no longer the Galilean or Poincaré group, but rather the Einstein group, that is, the group of four dimensional curvilinear coordinate transformations. It would therefore appear at first sight that the Einstein group should play a preferred role in the construction of a quantum theory of gravitation. However the Einstein group is generated by the constraints of the theory and must be eliminated in order to obtain the gauge-invariant observables of the theory appropriate for quantization [5].

Having thereby lost the preferred status of the space-time group, one could investigate the possibility of preserving some arbitrarily chosen sub-algebra of the classical observables as a set of linear mappings of the vector space. However, such a procedure is hardly unique, and is devoid of any semantic content. It is not clear what relation it would bear to quantization as we previously understood the procedure.

Rather than abandon all hope of a unique quantization of gravitation with a corresponding semantic identification of the observables, we can reimpose a preferred space-time group by the device of requiring restrictive boundary conditions on the class of solutions which we shall permit [6]. Such a program is indeed under investigation; however, further elaboration of it is not within the bounds of this paper.

References

1. NEUMANN, J. VON: Mathematical foundations of quantum mechanics. Princeton: Princeton University Press 1955.
2. UHLHORN, U.: Ark. Fys. **11**, 87 (1956).
3. EISENHART, L. P.: Riemannian geometry. Princeton: Princeton University Press 1926.
4. KOMAR, A.: Synthese **21**, 83 (1970).
5. BERGMANN, P. G., SCHILLER, R.: Phys. Rev. **89**, 4 (1953).
6. KOMAR, A.: Phys. Rev. **134**, B 1430 (1964).

V. Of Molecules and Galaxies

9 Algebraic Aspects of Formal Chemical Kinetics

RUTHERFORD ARIS [1]

Department of Chemical Engineering, University of Minnesota, Minneapolis, Minnesota

1 Introduction

It is striking how slight has been the impact of essentially mathematical ideas upon chemistry. Save in those reaches of the subject that are virtually identical with physics (e.g. quantum mechanics) the torch of foundations research has seldom been lit and the chemist has been far less aware than the physicist of the illumination that might be had from the concepts of rationality and structure that are the "euristic vision of mathematical trance" [1]. Such notions as the necessity and sufficiency of conditions, of similarity classes or of invariants wake but few echoes in the canyons of chemistry and attempts to explore the axiomatic strata that may be present in the chemical landscape have been rare. By comparison with physics, it has received scant attention from the philosophers of science, and, though such attentions are regarded by some as a mixed blessing, this has kept chemistry a little apart from the main intellectual current of natural philosophy. Thus, for example, there is no virtually no tract of it that rejoices in the clarity and coherence of the various physical theories that BUNGE has axiomatized and considered in his "Foundations of Physics" [2].

One approach to chemistry would be by way of the theory of interpenetrating continua which react with one another, for this is the logical extension of the ideas of continuum mechanics. There a coherent theory can be developed deductively from the concepts of bodies and their representations as manifolds, of physical space and its reference frames, of maps and families of fields representing place, density, force and stress. Work on these lines by TRUESDELL, BOWEN and others has progressed quite far and reveals the connection between stoicheiometry, thermodynamics and other physical principles, leading, for example, to the law of definite proportions as a theorem in BOWEN's analysis [3].

[1] "Acknowledgment is made to the donors of The Petroleum Research Fund, administered by the American Chemical Society, for partial support of this research."

There are however other aspects of chemistry that relate more to algebra and analysis than to field theory. Such is the question of molecular structure and its representation. It attracted the attention of the great J. J. SYLVESTER in a day when he could refer to "the *new* Atomic Theory" as "that sublime invention of KEKULE which stands to the *old* in somewhat similar relation as the Astronomy of KEPLER to PTOLOMEY'S or the System of Nature of DARWIN to that of LINNAEUS" [4]. His ideas, promulgated in prose that reflects at once the colour and vigour of his personality and the amplitude of an age more leisurely than ours, were in one way ahead of their time for, as WEYL points out [5], it required the discovery of quantum mechanics to reveal the significance of invariants. His theory even as extended by GORDAN and ALEXEJEFF failed to provide a physical interpretation of the addition of invariants or of the dynamical laws that would permit the calculation of binding forces and the chemist preferred to stick to his valence diagrams. More recently LEDERBERG [6] has used the classification of regular trivalent graphs and their association with certain polyhedra to provide a topological classification of chemical structure. Perhaps the deepest connection with algebra has been uncovered in the work of SELLERS [7] who has used the idea of a chain complex to show the structure of chemical mechanism. In this paper the chemist's notion of the mechanism of a chemical reaction will also be examined but the help of somewhat more elementary tools than those of SELLERS.

2 Stoicheiometry and Formal Chemical Kinetics

The word "*στοιχεῖον*" has an interesting history. Originally it meant one of a row or series and hence came to be used for a shadow line of a sundial, a letter of the alphabet or an element of knowledge. PLATO uses the word for the material elements – "For no one has yet explained the manner of their generation, but we speak of fire and the rest of them, whatever they mean, as though men knew their natures, and we maintain them to be first principles and letters or elements of the whole" [8]. The words sound very much as if he were prepared to treat them as we handle undefined primitives today, but the context shows that he is about to explain these things that "come into being through necessity" for he goes on to say that "they cannot reasonably be compared by a man of any sense to even to syllables or first compounds". However the word was to hand and the term "stoicheiometry" seems to have been minted by J. B. RICHTER in his "Anfangsgründe der Stöchiometrie, oder Meßkunst chemischer Elemente" (1792) and W. HAMILTON in his "Handbook of the Terms of the Arts and Sciences" (1825) speaks of stochiometry (sic) as "the Geometry of the chemical elements", a phrase of almost poetic insight [9]. By stoicheiometry we here mean the algebraical as much as the arithmetical or metrical aspects of the elements and the way in which they combine to form chemical species. Moreover we shall confine attention to the level of so-called empirical formulae in which alcohol is represented by C_2H_6O rather than C_2H_5OH, CH_3CH_2OH or any other partial representation of its structure.

In any particular universe of discourse [10] we recognize certain *elements* (denoted by their conventional symbols, such as C, H, O etc., or by E_k, according to the mood of the context) from which the chemical or *molecular species* that are present may be constructed. The formal similarity of C_2H_6O to $2C+6H+O$ shows that linear algebra is the natural language of stoicheiometry. Thus if the molecule M is constructed from ε_k atoms of E_k $(k=1, 2, \ldots, T)$ we can write its empirical formula as $\sum_{k=1}^{T} \varepsilon_k E_k$. If the molecular species $M_j, j=1, 2, \ldots, s$, take part in a reaction we can write this as a linear combination $\sum_{j=1}^{s} \alpha_j M_j=0$. For example if $E_1=H$, $E_2=O$, $M_1=H_2O$, $M_2=H_2$, $M_3=O_2$ we can write the reaction $2H_2+O_2 \rightarrow 2H_2O$ as $2M_1-2M_2-M_3=0$, i.e. $\alpha_1=2$, $\alpha_2=-2$, $\alpha_3=-1$.

The structure that we shall get by seeing what can be built from these elements may be called "formal chemical kinetics" since it pays no attention to the chemical nature of the substances but only to their formal representations. We shall first enquire in an algebraical way into what the chemist means when he explains a reaction by exhibiting its mechanism or the elementary steps by which it proceeds. When measurements of concentration are made and their rates of change are considered, we leave the province of algebra for that of analysis and the natural language is that of the theory of differential equations. The analytical aspect of the concept of mechanism can then be discussed.

3 *Aside on the Notions of Module and Morphism*

The notions of module and morphism that we shall need are illustrated in Fig. 1. Consider the collection of all complex integers, i.e. the numbers $m+in$, where m and n are integers, positive or negative. These elements can be represented by points (m, n) in the plane and they form an Abelian group under addition by the rule $(m+in)+(p+iq)=(m+p)+i(n+q)$. By this we mean that the sum is well defined and is an element, since $(m+p)$ and $(n+q)$ are integers, and has an inverse, since $(m+in)+(-m-in)=0+i0$ gives the identity element, $0+i0$. If we combine the set of complex integers with the ring of ordinary integers we have what is known as a module. The ordinary integers, which we will denote by Greek letters, form a ring since two operations, addition and multiplication, may be defined for them. If α and β are integers so also are $\alpha+\beta$ and $\alpha\beta$. Both operations have units, namely 0 and 1, since $\alpha+0=\alpha$ and $\alpha 1=\alpha$. However, only addition has an inverse, since $\alpha+(-\alpha)=0$ and $-\alpha$ is an integer, but, if $\alpha\beta=1$, β is not an integer. The combination of the group of complex integers $(m+in)$ and the ring of ordinary integers α is accomplished by

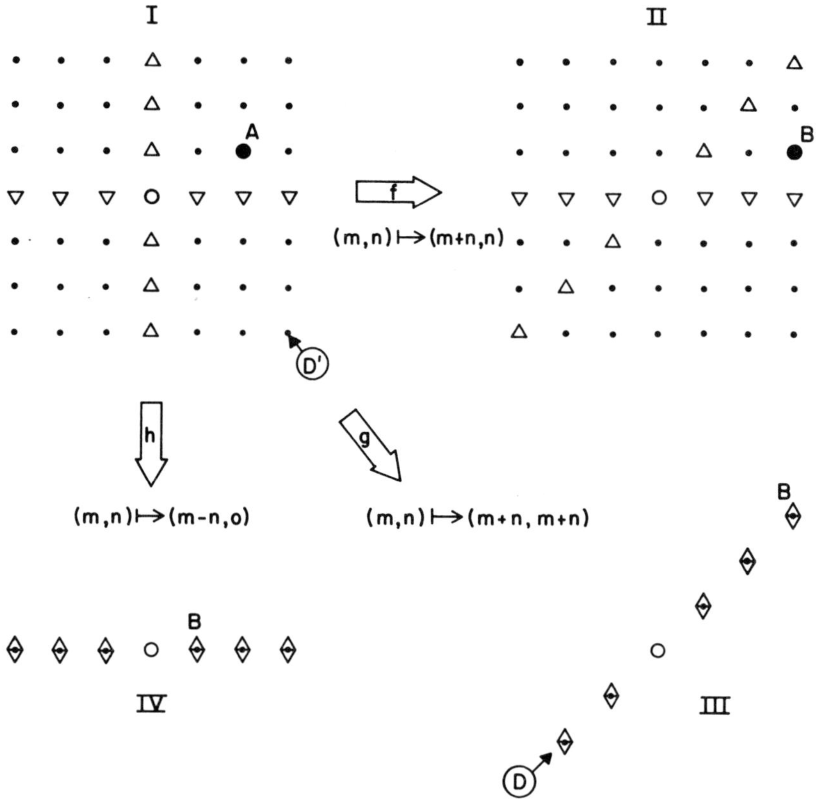

Fig. 1

defining the scalar multiplication $\alpha(m+in)$ to be the complex integer $\alpha m+i\alpha n$. This operation must also be distributive in the sense that if α and β are two integers from the ring and $l=m+in$ and $r=p+iq$ are two complex integers from the group, then $(\alpha+\beta)\,l=\alpha l+\beta l$, $(\alpha\beta)\,l=\alpha(\beta l)$ and $\alpha(l+m)=\alpha l+\alpha m$. All these rules are the formalization of everyday arithmetic. We can denote the algebraic entity thus obtained by M; it is a two dimensional free module since any element can be expressed in terms of two basic elements, $1+i0$ and $0+i1$.

Consider now the transformation under which the element $m+in$ is mapped into $(m+n)+in$ as shown by going from part I to part II of the figure. The typical point A in I is sent into B in II; the axis $(m+i0)$ remains the same but the axis $(0+in)$ becomes the 45° line (n, n). But the transformation is one-one, for each point in I has a unique image in II and each point of II is the image of a unique point in I. Moreover this transformation preserves the operation of addition of the complex numbers for $l \equiv (m+in)$ goes to $l' \equiv (m+n)+in$, $r \equiv (p+iq)$ goes to $r' \equiv (p+q)+iq$ and $l+r \equiv (m+in)+(p+iq)$ goes to $(m+p+n+q)+i(n+q)$, which is just $l'+r'$. In a similar way we see that scalar multiplication is preserved and this means that the transformation given by the rule $f(m+in)=(m+n)+in$ is a morphism of the module M onto itself. In particular it is an isomorphism since we have seen that it is one-one and the image of M under this morphism is identical with M itself, $fM \equiv M$. Compare this isomorphism with the morphism defined by $g(m+in)=(m+n)+i(m+n)$ which takes I into part III of the figure. Here the image of M lies entirely on the diagonal D of complex integers with equal real and imaginary parts. This is actually a submodule of M for all the complex integers of the form $m+im$ have all the properties of a module, yet they are only a subset of the elements of M. In contrast to the first morphism where only the zero element $0+i0$ was mapped into the zero element we now have all the elements on the negative diagonal D' of I mapped on the origin, for $g(m-im)=(0+i0)$. The set of elements, D', which are sent into the zero under the morphism g, is called the kernel of g or Ker g. On the other hand the set D of elements which are images of elements of M is called Im g. The relation between parts I and IV of the figure illustrates another morphism defined by $h(m+in)=(m-n)+i0$. This time Im h is the set of real integers, R, and Ker h is the diagonal set D of integers $m+im$.

We observe that if the transformation g is followed by h, $(m+in)$ first goes to $(m+n)+i(m+n)$ and then to $0+i0$. Thus $hg(m+in)=0+i0$ and the whole module has been transformed into the trivial submodule consisting of the zero $0+i0$. The sequence of morphisms g, h is said to be exact if it has this property, namely that Im $g=$ Ker h.

4 *Chemical Mechanism: The Algebraic Aspect*

The algebraical theory which has just been outlined is extraordinarily reminiscent of the business of balancing chemical reactions with respect to each element. The reaction $2H_2O-2H_2-O_2=0$ is a proper chemical reaction because the transformation, or linear combination, by which it is formed from the molecular species H_2O, H_2, O_2 bears such a relation to the transformation by which they are formed from the atomic species

H and O that the net number of atoms of each element entering the reaction is zero. The species of interest are very much like the image of some structural transformation by which they are composed from their atoms and they are to form the kernel of a transformation by which reactions are formed from them. Consider for instance a module whose elements are all of the form $\begin{bmatrix} m & \mathrm{H} \\ m & \mathrm{O} \end{bmatrix}$ where m is an integer, positive or negative. This is a module over the ring of integers for we may define

$$\begin{bmatrix} m & \mathrm{H} \\ m & \mathrm{O} \end{bmatrix} + \begin{bmatrix} n & \mathrm{H} \\ n & \mathrm{O} \end{bmatrix} = \begin{bmatrix} (m+n) & \mathrm{H} \\ (m+n) & \mathrm{O} \end{bmatrix}$$

and

$$v \begin{bmatrix} m & \mathrm{H} \\ m & \mathrm{O} \end{bmatrix} = \begin{bmatrix} v\,m & \mathrm{H} \\ v\,m & \mathrm{O} \end{bmatrix}$$

where m, n and v are any integers. In fact this is very like the submodule D of diagonal elements in the earlier figure. A similar module of molecular species could be formed from the elements

$$\begin{bmatrix} p & \mathrm{H_2O} \\ p & \mathrm{H_2} \\ p & \mathrm{O_2} \end{bmatrix}$$

where p is an integer. There is clearly a morphism of the module of atomic species, $\mathfrak{E}$, into that of molecular species, $\mathfrak{M}$, which can be conveniently represented in matrix notation by

$$\begin{bmatrix} p & \mathrm{H_2O} \\ p & \mathrm{H_2} \\ p & \mathrm{O_2} \end{bmatrix} = \begin{bmatrix} 2 & 1 \\ 2 & \cdot \\ \cdot & 2 \end{bmatrix} \begin{bmatrix} p & \mathrm{H} \\ p & \mathrm{O} \end{bmatrix}.$$

A reaction is a linear combination of species such as $l\,\mathrm{H_2O} + m\,\mathrm{H_2} + n\,\mathrm{O_2}$. Such linear combinations also form a module, say $\mathfrak{R}$, for we can define a sum by

$$\begin{aligned}(l\,\mathrm{H_2O} + m\,\mathrm{H_2} + n\,\mathrm{O_2}) &+ (l'\,\mathrm{H_2O} + m'\,\mathrm{H_2} + n'\,\mathrm{O_2}) \\ &= (l+l')\,\mathrm{H_2O} + (m+m')\,\mathrm{H_2} + (n+n')\,\mathrm{O_2}\end{aligned}$$

and a scalar multiplication by

$$v(l\,\mathrm{H_2O} + m\,\mathrm{H_2} + n\,\mathrm{O_2}) = v\,l\,\mathrm{H_2O} + v\,m\,\mathrm{H_2} + v\,n\,\mathrm{O_2}.$$

For a reaction to be proper it must be balanced in both H and O (or $l{:}m{:}n = 2{:}-2{:}-1$) and hence the morphism from the module $\mathfrak{M}$ into the reaction module $\mathfrak{R}$ must map the species onto the zero of $\mathfrak{R}$ if it represents the proper reactions that may take place between the species. Calling the first morphism β and the second α we may say that the sequence

$$\mathfrak{E} \xrightarrow{\beta} \mathfrak{M} \xrightarrow{\alpha} \mathfrak{R}$$

is exact at $\mathfrak{M}$ if the reactions "α" are proper for the structure "β". If we add at the two ends a morphism from and to the trivial module 0 we have a "short exact sequence" beloved of algebraists,

$$0 \to \mathfrak{E} \to \mathfrak{M} \to \mathfrak{R} \to 0.$$

In this sequence $0 \to \mathfrak{E}$ means that the trivial module is mapped on to the zero of $\mathfrak{E}$ and $\mathfrak{R} \to 0$ means that every element of the module $\mathfrak{R}$ is mapped onto the trivial module. Thus the short exact sequence is exact at $\mathfrak{E}$, $\mathfrak{M}$ and $\mathfrak{R}$.

We observe that the "elements" in the algebraic representation need not be elements in the chemical sense. For example in the reaction represented by

$$A + B \rightleftharpoons P,$$

the substance A might be carbon dioxide and B might be water, P being carbonic acid. We do not need to express the molecular species in terms of carbon, hydrogen and oxygen for quite clearly P must have A and B as constituent parts. A and B can therefore be treated as "elements" and we have the structural transformation represented in matrix form by

$$\begin{bmatrix} A \\ B \\ P \end{bmatrix} = \begin{bmatrix} 1 & \cdot \\ \cdot & 1 \\ 1 & 1 \end{bmatrix} \begin{bmatrix} A \\ B \end{bmatrix}.$$

For this system the only possible reaction is a multiple of $-A-B+P=0$. A mechanistic explanation of such a reaction might involve an enzyme E (in the above case carbonic anhydrase activated by a zinc ion) with which the reactant A forms a complex $C \equiv (AE)$. This then reacts with B to form P and gives back the enzyme. Thus the reaction

$$R \equiv -A - B + P = 0$$

is held to be explained by the sequence

$$\begin{aligned} S &\equiv -A \quad \cdot \quad \cdot - E + C = 0 \\ T &\equiv \quad \cdot - B + P + E - C = 0. \end{aligned}$$

It is clear that stoicheiometrically the overall reaction is the sum of these two reactions. We now have three elements A, B, E in which the reactions must be balanced and five species A, B, P, E, C take part in them. The morphism from the enlarged module $\mathfrak{E}^+$ that we can form from A, B

and E to the enlarged module $\mathfrak{M}^+$ formed from the five species is represented by

$$\begin{bmatrix} A \\ B \\ P \\ E \\ C \end{bmatrix} = \begin{bmatrix} 1 & \cdot & \cdot \\ \cdot & 1 & \cdot \\ 1 & 1 & \cdot \\ \cdot & \cdot & 1 \\ 1 & \cdot & 1 \end{bmatrix} \begin{bmatrix} A \\ B \\ E \end{bmatrix}.$$

The morphism from $\mathfrak{M}^+$ to $\mathfrak{R}^+$, the enlarged reaction module, must give balanced reactions and in fact there can only be two independent reactions, which are linearly dependent on S and T. The overall reaction which the mechanism explains is the only linear combination of S and T that omits all reference to E and C, namely the reaction R. Since the reactions $\mathfrak{R}^+$ are proper for the species $\mathfrak{M}^+$ the sequence

$$0 \rightarrow \mathfrak{E}^+ \rightarrow \mathfrak{M}^+ \rightarrow \mathfrak{R}^+ \rightarrow 0$$

is a short exact sequence.

Moreover we can see obvious projections of $\mathfrak{E}^+$ and $\mathfrak{M}^+$ into $\mathfrak{E}$ and $\mathfrak{M}$ respectively, namely the mappings

$$\lambda: \begin{bmatrix} A \\ B \\ E \end{bmatrix} \mapsto \begin{bmatrix} A \\ B \end{bmatrix} \qquad \mu: \begin{bmatrix} A \\ B \\ P \\ E \\ C \end{bmatrix} \mapsto \begin{bmatrix} A \\ B \\ P \end{bmatrix}.$$

The mapping from $\mathfrak{R}^+$ to $\mathfrak{R}$ is

$$\nu: \begin{bmatrix} S \\ T \end{bmatrix} \mapsto [R], \qquad R = S + T.$$

The coefficients in this mapping, in this case (1, 1), are known as the stoichiometric numbers. Thus the algebraist would say that the chemical kineticist is concerned to find a parallel sequence to explain a given reaction which will form the commutative diagram

$$\begin{array}{ccccccccc} 0 \rightarrow & \mathfrak{E}^+ & \xrightarrow{\beta^+} & \mathfrak{M}^+ & \xrightarrow{\alpha^+} & \mathfrak{R}^+ & \rightarrow 0 \\ & \downarrow{\scriptstyle\lambda} & & \downarrow{\scriptstyle\mu} & & \downarrow{\scriptstyle\nu} & \\ 0 \rightarrow & \mathfrak{E} & \xrightarrow{\beta} & \mathfrak{M} & \xrightarrow{\alpha} & \mathfrak{R} & \rightarrow 0. \end{array}$$

This opens up the possibility of algebraically describing all the possible mechanisms which are consistent with a given enlargement of $\mathfrak{E}$ and $\mathfrak{M}$ to $\mathfrak{E}^+$ and $\mathfrak{M}^+$. However there is a limitation to the usefulness of this, for though we can determine the morphism α^+ that is exact in conjunction with β^+ we can only determine its matrix representation up to a non-singular linear transformation. In a complicated problem it is not clear how one would display the various forms that reaction scheme might take so as to allow the chemist to pass judgement on those worthy of investigation.

5 Beyond Stoicheiometry

Stoicheiometry in chemical kinetics plays somewhat the same role as does kinematics in continuum mechanics, it defines the "motions" that are compatible with the basic postulates. Just as in mechanics we then look to the dynamical principles and constitutive relations to give us equations that govern the rate at which a possible motion takes place, so also in chemistry we pass from the algebra of stoicheiometry to the analysis – specifically, the theory of differential equations – of kinetics. The role of the principle of conservation of momentum is played by the assumption that in a uniform closed system the rates of change in the amount or concentration of each species can be expressed as functions of these concentrations. The particular form that these functions may take corresponds to the constitutive relation in mechanics. In the reaction $R: -A-B+P=0$ considered above the rate of formation of P might be given by

$$\frac{dp}{dt}=\frac{Kab}{1+K_1 a+K_2 b} \tag{$*$}$$

where lower case letters denote corresponding concentrations and K, K_1, K_2 are constants. Such an expression may be useful in that it fits the experimental data in a given situation, but the chemist would regard it as remote from the stoicheiometry of the reaction and so as requiring some "explanation" in terms of mechanism. If the reactions S and T are taken as the elementary steps, then

$$-\frac{da}{dt}=k_1 a e-k_2 c\colon \frac{dp}{dt}=k_3 b c.$$

This pair of equations can be reconciled with the previous one by the theory of singular perturbations of differential equations [11]. Let us write the equations for the rates of change of the concentrations of A, B and the complex C and make use of the fact that the total concentration of enzyme is constant (e_0) so that $e=e_0-c$;

$$\begin{aligned}\dot{a}&=-k_1 a(e_0-c)+k_2 c\\ \dot{b}&=-k_3 b c\\ \dot{c}&=k_1 a(e_0-c)-k_2 c-k_3 b c.\end{aligned}$$

A natural unit for the concentration of c is e_0, the total enzyme concentration, just as that for a, b or p is a_0, the initial concentration of A. Setting $x=a/a_0$, $y=b/a_0$, $z=c/e_0$ we have

$$\frac{1}{k_1 e_0}\frac{dx}{dt} = -x(1-z)+\left(\frac{k_2}{k_1 a_0}\right) z$$

$$\frac{1}{k_1 e_0}\frac{dy}{dt} = \qquad\qquad -\left(\frac{k_3}{k_1}\right) y z$$

$$\frac{1}{k_1 a_0}\frac{dz}{dt} = +x(1-z)-\left(\frac{k_2}{k_1 a_0}\right) z-\left(\frac{k_3}{k_1}\right) y z.$$

Now what is notable about these equations is that the natural dimensionless time in the first two is $k_1 e_0 t$ and in the last is $k_1 a_0 t$ and, if $a_0 \gg e_0$, time is, so to speak, running very much faster for the last equation than it is for the others. Thus x and y change relatively slowly and z is able to come to its steady state with respect to the current values of x and y, namely

$$z=k_1 a_0 x/(k_1 a_0 x+k_2+k_3 a_0 y)$$

or

$$c=e_0 k_1 a/(k_1 a+k_2+k_3 b).$$

But substituting this in the equation for the rate of formation of the product we have

$$\frac{dp}{dt}=k_3 b c=\frac{e_0 k_1 k_3 a b}{k_1 a+k_2+k_3 b}=\frac{e_0(k_1 k_3/k_2) a b}{1+(k_1/k_2) a+(k_3/k_2) b}.$$

This is the same as the observed rate equation, given above as (*), if we take $K=e_0 k_1 k_3/k_2$, $K_1=k_1/k_2$ and $K_2=k_3/k_2$. Thus the hypothesis $a_0 \gg e_0$ taken to the extreme $(e_0/a_0)\to 0$ fuses the three rate equations into one which, if it be the observed empirical rate law, is thus explained in analytical, as well as algebraic, terms by the mechanism $R=S+T$.

Algebraic considerations are also involved in the forms that the reaction rate expressions may take. If they are linear in the concentrations then the whole apparatus of linear algebra may be invoked and some remarkable results obtained as Wei and Prater [12] have shown. For higher orders of reaction the algebraic theory is much more recondite and the structural picture less complete [13].

6 Postscript

We have only been able to touch here on a few of the algebraic and analytical ideas that are latent in chemical kinetics [14]. Much remains to be done to exploit them in examining the foundations of chemistry.

In his "Preliminary Discourse on the Study of Natural Philosophy" (London, 1830), JOHN HERSCHEL looked forward to the day when the experimental phase of chemistry would be succeeded by a period in which it would become a deductive, mathematical science. If we cannot take his somewhat supercilious view that chemical theories are "for the most part of that generally intelligible and readily applicable kind, which demand no intense concentration of thought, and lead to no profound mathematical researches," at least we need not share his pessimistic feeling that "the axioms of chemistry, the true handles of deductive reasoning, ... (are) still unknown, and, perhaps, likely long to remain so."

References

1. BRIDGES, ROBERT: The testament of beauty Book I. l. 368.
2. Springer Tracts in Natural Philosophy. Vol. 10. Berlin-Heidelberg-New York: Springer 1967.
3. BOWEN, R. M.: Arch. Rational Mech. Anal. **29**, 114 (1968). See also KRAMBECK, F. J.: ibid. **38**, 317 (1970). The work of M. FEINBERG, soon to be published in the same journal, combines some aspects of Bowen's approach with deeper notions of linear algebra.
4. SYLVESTER, J. J.: On an application of the new atomic theory to the graphical representation of the invariants and covariants of binary quantics. Amer. J. Math. **1**, 64–125 (1878).
5. WEYL, H.: Philosophy of mathematics and natural science, 2nd edn. Princeton University Press 1949. Appendix D.
6. LEDERBERG, J.: Proc. Nat. Acad. Sci. U.S.A. **53**, 134 (1965).
7. SELLERS, P. H.: SIAM J. Appl. Math. **15**, 13 (1967). For an introduction to this paper see an exposition by SELLERS in J. Franklin Inst. **290**, 113 (1970), and brief mention on pp. 19 and 20 of the review [14]. A further article by SELLERS will appear in Arch. Rational Mech. Anal. in late 1971 or soon after.
8. Tim. 48c. JOWETT's Translation. *«νῦν γαρ οὐδεῖς πω γενεσιν αὐτων μεμήνυκεν, ἀλλ' ὡς εἰδοσιν πῦρ ὅτι ποτέ ἐστιν και ἕκαστον αὐτῶν λέγομεν ἀρχας αὐτὰ τιθέμενοι στοιχεῖα του παντός.»*
9. The use has apparently been confined to chemistry, though TYNDALL in his treatise on Heat (1880) refers to "the doctrine of the conservation of force, or, as I should express it, Physical Stoichiometry".
10. Thus carbon, hydrogen and oxygen suffice for large tracts of organic chemistry, as BRIDGES recognized when he wrote:
    "... whether it be starch, oil, sugar or alcohol
    'tis ever our old customers, carbon and hydrogen,
    pirouetting with oxygen in their morris antics"
    The Testament of Beauty Book III. ll. 935–7.
11. The appropriateness of singular perturbation theory was first discovered by A. ACRIVOS, J. R. BOWEN and A. K. OPENHEIM [Chem. Engng. Sci. **18**, 177 (1963)]. For a discussion of the enzyme reaction see F. G. HEINEKEN, H. M. TSUCHIYA and R. ARIS, Math. Biosci. **1**, 95 (1967).
12. PRATER, C. D., WEI, J., in: Advances in catalysis, vol. 13. New York: Acad. Press 1962.
13. See for example ARIS, R.: Ind. Eng. Chem. Fundamentals **3**, 28 (1964) and WEI, J.: Ind. Eng. Chem. Fundamentals **4**, 61 (1965).
14. For a more general, but still only partial, review see ARIS, R.: Mathematical aspects of chemical reaction. Ind. Eng. Chem. **61**, 17 (1969).

VI. Philosophical

10 Criteria for Meaning Changes in Theoretical Physics

SCOTT A. KLEINER

Department of Philosophy, University of Georgia
Athens, Ga.

A theory can be revised by replacing or modifying one or more of its component statements. Such revision is usually thought of as and often is of a straightforward factual claim. The discovery of a tenth planet would require simply that one specification of the solar system's components be replaced by another in the classical mechanical theory of the solar system's motion. The concepts used in this theory, e.g. 'mass', 'force', 'natural satellite', seemingly suffer no revision as a consequence of this type of theoretical revision.

Recently a number of philosophers have called our attention to what they claim to be more far-reaching types of theoretical revision. Not only the factual content but also the very meaning of the expressions used in a theory are subject to revision. The theoretician may introduce new concepts, that is new expressions with new meaning, as well as revise concepts already in use.

Some philosophers have claimed or implicitly assumed that all theoretical revisions amount to conceptual changes [17, 8, 12]. The rejection of any distinction between conceptual and factual change has led them to reject a number of tenets central to what is generally called empiricism and realism. These tenets include the belief that observational evidence is methodologically relevant to the acceptance and rejection of theories, that distinct theories can be compared in regard to their relative value, and that distinct theories can be formulated about the same objects.[2]

I have no doubt that there are episodes in the history of science in which new concepts are introduced or old expressions are given new meanings, and new expressions with new meanings are used profitably for the first time. Yet I believe that criteria for diagnosing conceptual

[1] I would like to express my gratitude to Professor MARIO BUNGE for his helpful comments on an earlier draft of this paper.

[2] For a more thorough account of the consequences of these views, see [16, 2, 1, 9, 7].

changes and the logical and methodological import of such changes have yet to be adequately specified. In what follows I will attempt to (1) specify what does and what does not serve as a useful indicator or criterion of conceptual change (2) show that conceptual change can be and often is a matter of gradual evolution via discrete episodes rather than complete replacement (3) make, clarify and defend the claim that conceptual change is a change (not necessarily an improvement) in the capacities of individuals or groups for conducting research in contrast to a change in the results of exercising these capacities. I believe that my contribution to the completion of these tasks should help to appropriately clarify and qualify the above tenets of empiricism, which I regard as incontrovertible but unclear.

1 Criteria for Meaning and for Meaning-Change

One is tempted to say that episodes in which the concepts of a given field are revised or completely replaced can be identified by the introduction of symbols hitherto not appearing in sentences expressing observations and theories of the field. However, the possibility of ambiguity or equivocation make the occurrence of specific symbols insufficient for determining the conceptual apparatus in use at a given stage in the history of science [17], pp. 101 f; [8], p. 59; [13].

Semantical rules, e.g., '"*F*" designates force,' '"*p*" can be used to refer to one of the set of protons *P*', are also of little use in diagnosing conceptual selection or change. The rules cannot be understood unless the observer of scientific development appreciates *the* conception of force and proton intended in these statements. However, the possibility of equivocation applies equally to expressions used in stating semantical rules as to expressions used in stating the laws and principles of a theory. 'Force' in the rule might have any of the several meanings that '*F*' might have in the several actual theories using this expression.

The above-mentioned semantical rules are generally taken as conventional metatheoretical statements, for the expression-types mentioned (and not used) in them (*F*, *p*, etc.) are arbitrarily selected for correlation with extra-theoretical objects or concepts. Some writers have distinguished *semantical assumptions* from semantical rules, claiming that the assumptions are not conventional but are proper hypotheses that might be false (cf. [4], pp. 18–34; [3], Chapter 2; [5], pp. 467–471). I can concede this distinction with this proviso: In *theoretical frameworks* such as classical particle or quantum mechanics or relativistic particle mechanics, the semantical assumptions are general existential and not singular statements.

For example, classical particle mechanics contains as a proper axiom the equivalent sentences $\Sigma \neq \Lambda$ or $(\exists x)(x \in \Sigma)$, which, in view of semantical rules for 'Σ', can be interpreted as 'The set of material particles is not empty,' or 'There exists a material object.' Similarly, fundamental particle theories should contain '$P \neq \Lambda$' or its equivalent as a proper axiom. The latter axiom allows us to take the rule '"p_i" ($i = 1, \ldots, n$) can be used to refer to a proton' as granting permission for the referential use of n individual names 'p_i' ($i = 1, \ldots, n$), or of the singular descriptions $\iota x (x \in P \& \ldots i \ldots)$, where the $\ldots i \ldots$ is to be filled with a singular proposition specifying the spatiotemporal extent of the i-th particle p_i. (Note that the rule as stated is incomplete, for it does not specify exact circumstances under which 'p_1', 'p_2', etc. do in fact refer.)

Nevertheless, both the semantical rule and the assumption it presupposes can be used equivocally, and hence their appearance among axioms and metastatements of two different theories is not sufficient to justify the claim that 'P' and 'p_i' mean the same in their occurrences in the two theories.

It is widely held that a necessary condition for two expressions having the same meaning is that one expression can be substituted for one other in a logically or analytically true sentence so as to preserve the necessary truth of that sentence. Accordingly, if the necessary truth of a sentence is not preserved under such substitution, we may conclude that the expressions used in the substitution differ in meaning. However, this relation between sameness of meaning and substitutivity cannot be used as a criterion for meaning change in scientific development because substitutivity is one of the questions at issue in determining how two different theories might be compared. That is, in comparing two theories, T_1 and T_2, an item of interest is whether statements and arguments in T_1 can be paraphrased by statements and arguments in T_2. In other words, can expressions in T_1 be substituted for similar expressions in statements and arguments in T_2 without altering the validity of these arguments or the truth of these statements? Substitutivity as a criterion for meaning change simply begs this important question.

The syntax of a theory's non-logical expressions is relevant to but insufficient for determining uniquely the concepts designated by these expressions, as we shall see below. One theory might construe 'mass' as a property, and another might construe it as a relation. However, any particular theory might mention several properties or relations, so that within a theory specification of an expression's syntactical properties leaves open the possibility of equivocation. *A fortiori*, if the expression 'M' is characterized as a predicate in one theory and a relation in the other, it does not follow that the same concept 'M' is revised in going from one to the other theory. 'M' might be used to designate mass in one and force in the other theory, two completely distinct concepts [13].

Semantical rules of the kind so far considered are usually advanced in order to specify the *extension*, *denotation* or *reference* of the non-logical expressions used in a theory. However, most students of semantics are aware that these concepts do not capture all of an expression's 'meaning,' for expressions can be co-extensive or co-referring and yet 'mean' quite different things, e.g., 'the author of *Waverly*' and 'Sir Walter Scott,' or 'rational animal' and 'extant featherless biped.' Accordingly, it is commonplace to distinguish an expression's *sense* or *intension* from its reference or extension (cf. [10]). Meaning then is construed as determined by both the sense and reference of an expression in a manner expressed by, e.g., the formula $M(X, CM)=\langle S(X, CM), R(X, CM)\rangle$, where '$M(X, CM)$' abbreviates 'the meaning of expression X in classical mechanics,' '$S(X, CM)$' is 'the sense of X in classical mechanics,' and '$R(X, CM)$' is 'the reference of X in classical mechanics.'[1]

In what follows I will be concerned with specifying those components (statements or hypotheses) of a theory that determine the sense of that theory's non-logical expressions. A change in these components is necessary and sufficient for a shift in the sense of the relevant expression, and consequently such a change is sufficient for a shift in that expression's meaning. (I have discussed criteria for referring potential and actual reference elsewhere [16].)

I propose that a satisfactorily enriched statement of any given theory should be accompanied with statements the understanding and assent to which is necessary and sufficient for ascribing capacities for conducting research within that framework. These statements determine the sense of the extralogical expressions used in that framework. In the case studies below, I will try to show that these statements, criterial statements, provide logical links between these and other like expressions and between these expressions and expressions already in use either in other theories or in ordinary language supported by a background of commonsense beliefs.

The criterial statements can, but need not, specify both conditions fully justifying a theoretical term's application (sufficient conditions) and conditions the failure of which fully justifies the denial of its application (necessary conditions). That is, a criterial statement might be phrased 'For all x, x has mass if and only if $P_1 x$ and $P_2 x$ and ... and $P_n x$.' ($P_1, \ldots, P_n$ stand for the specified criteria.) It might also, but need not, specify conditions that are jointly necessary but individually sufficient. That is, the criterion might be phrased 'For all x, x has mass if $P_1 x$ or $P_2 x$ or ... or $P_n x$, and x has mass only if $P_1 x$ and $P_2 x$ and ... and $P_n x$.' Further alternatives might include the specification of any number of the $P_1, \ldots, P_n$ as individ-

[1] The advantages of so construing meaning were pointed out to me by Prof. MARIO BUNGE in a personal communication.

ually or jointly necessary and any number of the same list, and not necessarily including any necessary conditions, as individually or jointly sufficient conditions. For example, our criterion might be 'For all x, x has mass if $P_1\,x$ and $P_2\,x$ and only if $P_4\,x$ or $P_5\,x$ or $P_6\,x$'.

The properties $P_1, \ldots, P_n$ need not be directly observable. That is, they need not be properties for which there are epistemic standards of the form '*a* would not look (taste, feel, smell, sound, etc.) *L* under conditions *C* unless it were P_i' [7]. A property P_i could also be linked to directly observable properties by physical law, or for the present there may be no known laws linking P_i to such properties.

2 An Example: Criteria for "... has Mass"

That an object is seen, felt, heard, tasted or smelled are individually sufficient conditions for ascribing mass. In fact, the weaker conditions of visibility, tangibility, etc., are sufficient conditions for attributing that quality. The list of alternative sufficient conditions is extended in physics to include properties of microphysical objects that cannot be directly observed, e.g., a macrophysical object's being capable of capture or bottling (Lavoisier's or Hero of Alexandria's 'gases' or 'airs') a microphysical object's possessing weight or possessing less than maximal velocity of light in relativistic kinematics (cf. [21], especially pp. 199f., 208–222).

The last mentioned of the above conditions is necessary for an object's having classical or relativistic mass. However, I believe that this condition was not widely recognized in classical mechanics before the advent of the special theory of relativity, for there were surely pre-relativistic physicists willing to contemplate infinite velocities generated by infinite forces in finite masses or infinite velocities in infinite time, generated by finite forces. Rather, most of the various attempts to 'define' classical mass take the capacity for exerting an impulse or (equivalently) for transferring momentum as a necessary condition for the possession of mass (cf. [14], Chapter 8). This aspect of the concept of mass was revised after the relativistic account of radiation pressure appeared.

The sufficient conditions mentioned above are always *logically relevant*[1] but are *useful* for deciding application of 'has mass' under some but not under all circumstances. Massive objects are visible only under certain conditions, and many of these conditions are common knowledge that can be explained by the laws of wave-optics and the anatomy and physiology of our visual receptors. In the light of this knowledge, an object's having been seen is a useful criterion for the application of

[1] Any condition which is by itself and not by appeal to regularities or laws of nature either necessary or sufficient for applying a term is logically relevant to its application.

'... has mass' if conditions of visibility exist or can be produced artifically. Our technical shortcomings can affect the utility but not the logical relevance of a criterion. Accordingly, if we cannot weigh electrons, weight is not a useful sufficient condition for ascribing mass to them. However, the invention of the appropriate instrument or techniques or the formulation or application of laws relating electron mass to technically observable properties can make the logically relevant quality weight a useful sufficient condition for calling electrons material objects.

The more extended a list of logically relevant conditions we have, the wider the range of objects that fall into the scope of our activities of ascribing mass. Also, the greater our awareness of the utility of criteria already known to be logically relevant, the more deliberate will be our actions of determining whether mass ought to be ascribed. That is, we may select in advance criteria known to be useful rather than randomly trying to evaluate all properties, whether logically relevant or useful. Accordingly, one can sharpen his conception of mass by acquiring knowledge of the community's technological capacities, and the community can extend its capabilities in attributing mass by advancing the technology relevant to evaluating properties logically relevant to 'has mass'.

Note that on this analysis we can conceive of mass in domains in which we lack the technical capacity or useful criteria for its application. Thus, 'A proton at the sun's center has mass' is a cognitively meaningful sentence.

3 Background and Theory-Specific Criteria

In the preceding section criteria for 'has mass' have been stated in two different ways. They are phrased in general and perhaps vague terms including 'visible', 'tangible', and 'incapable of maximal velocity'. These we might call statements of *background criteria*. They are also stated in forms making them specific to the theories mentioned, pre-classical, classical and relativistic mechanics. The latter statements are patently less general and perhaps less vague, and might be called *theory specific criteria*. The distinction between these two types of criteria help us to understand several features of conceptual development, whether in the individual or in the community.

In the first place, it is in the light of background criteria and shared beliefs or conceptions that we may determine whether two theories are mutually *contradictories*, *contraries* or *subcontraries*. Theories T and T' are contradictories if the truth of the first entails the falsity of the second, the falsity of the first entails the truth of the second, and, conversely, the falsity of the second entails the truth of the first and the truth of the second entails the falsity of the first. T and T' are contraries if the truth of the first entails the falsity of the second and the truth of the second entails the falsity of the first, but the falsity of neither entails the truth of the other. T and T' are subcontraries if the falsity of the first entails the truth of the second and the falsity of the second entails the truth of the first, but the truth of neither entails the falsity of the other.

Let us consider two scientists, S_1 and S_2, who agree that they are studying the same object, a.[1] Suppose also that S_1 believes that this object has mass, and his belief can be expressed by the sentence $M^1\, a$. Also, S_2 believes that it is not the case that a has mass, or $-M^2\, a$. (The superscripts here and in what follows identify the vocabulary, S_1's or S_2's vocabulary, in which the predicate occurs.) Furthermore, S_1 subscribes to criteria which can be stated thus: P_1^1 is sufficient for the application of M^1, or $(x)\,(P_1^1\, x \Rightarrow M^1\, x)$, and P_2^1 is necessary for M^1's application, or $(x)(-P_2^1\, x \Rightarrow -M^1\, x)$. S_2 subscribes to criteria stated as $(x)(P_1^2\, x \Rightarrow M^2\, x)$ and $(x)(-P_2^2\, x \Rightarrow -M^2\, x)$.[2] On the basis of these criteria the following cases provide sufficient, and in some cases, minimal conditions for S_1's and S_2's theories being (1) complementary, (2) sub-contrary, (3) contrary and (4) contradictory:

1. $M^1\, a$ and $-M^2\, a$ are complementary on the basis of the above criteria if the conjoined statements '$P_1^1\, a \;\&\; -P_2^2\, a$' and '$-P_2^1 \;\&\; P_1^2\, a$' each represent a possible state of affairs. Under the stated criteria, the first statement entails that both $M^1\, a$ and $-M^2\, a$ are true and the second entails that they both are false. Note also, if '$P_1^1\, a$ or $-P_2^1\, a$ or $P_1^2\, a$ or $-P_2^2\, a$' does not represent a technically or physically possible state of affairs, S_1's and S_2's theories are *methodologically incommensurable* with respect to the given criteria. That is, the criteria link their theories with

---

[1] 'a' is a singular term used by both S_1 and S_2 for referring or identifying individual objects of study. Demonstratives such as 'this cloud chamber', proper names such as 'Jenny' and singular descriptions such as 'the cloud chamber in room 202 in ... (building), ... (institution), etc.' all are singular terms. If 'a' and 'b' are distinct singular terms referring to distinct objects bearing to one another no relations such as composition (a is part of b), participation (object a participates in event b), or other similar *(ontic)* relations, then '$M^1\, a$' and '$M^2\, b$' are complementary.

[2] Here '(x)' stands for the usual quantifier 'for all x' and '$\Rightarrow$' is the propositional connective 'if ... then ...'. One might object at this point that my account of meaning change is question begging. It appears that in attempting to explain meaning change I have already taken it for granted that we already understand what it is for two properties to be identical or distinct, or in other terms, what it is for two predicates to be alike or different in meaning. However, my concern is to elucidate likeness, diversity and change in concepts introduced by physicists for the purpose of theorizing in physics, that is, likeness, diversity and change in technical terms. The properties P_1^1 might themselves be technical, but in that case identity, etc., should be explained by linking these to further properties, say $Q_1^1, \ldots, Q_n^1$. Ultimately we should arrive at a set of criteria mentioning commonplace or non-technical notions. Then to carry out our task all we need is to be able to decide practically identity, etc., among these non-technical concepts, an ability we have developed from childhood.

no possible states of affairs that would warrant the acceptance or rejection of one in the light of some or no evidence for or against the other.

2. S_1's and S_2's claims are sub-contraries if and only if for any object a, $P_2^2\, a$ and $-P_1^1\, a$ cannot both be true. Now, if S_1 and S_2 share the belief that $P_2^2\, a$ implies $P_1^1\, a$, then either $M^1\, a$, or $-M^2\, a$, or perhaps both, are true. *A fortiori*, if S_1 and S_2 conceive P_1^1 and P_2^2 in their background beliefs as the same property, their theories are subcontraries.

3. $M^1\, a$ and $-M^2\, a$ are contraries if S_1 and S_2 agree at least that $P_2^1\, a$ implies $P_1^2\, a$. *A fortiori*, if they identify these properties in their background beliefs, both of their theories cannot be true.

4. If S_1 and S_2 agree that $P_2^2\, a$ implies $P_1^1\, a$ and $P_2^1\, a$ implies $P_1^2\, a$, then $M^1\, a$ and $-M^2\, a$ cannot both be true and cannot both be false. These are minimal conditions for the two theories being contradictories. Again, the stronger condition that P_1^1 and P_2^2, and P_1^2 and P_2^1 are respectively identical also entails that the two theories are contradictories.[1]

We can also appeal to the distinction between theory-specific and background criteria for clarification of the commonplace claim that a new theory can further clarify a concept already current is scientific or everyday thought. Classical continuum and particle mechanics contains considerable elaboration and refinement of the pre-classical 'change in condition,' the effects of forces. Relativistic particle mechanics offered an alternative specification to the classical notion of maximal velocity. Insofar as puzzles about infinite forces and velocities were unresolved in classical particle mechanics, relativity can be viewed as containing a positive contribution to the clarity of this concept and the related concept of mass.

The theory-specific criteria stand and fall with the theory to which they are characteristic and peculiar. The change or the further specifica-

---

[1] The following arguments are used in the analysis of these cases:

Case 1:

$$\begin{array}{l} (x)(P_1^1\, x \Rightarrow M^1\, x) \\ (x)(-P_2^2\, x \Rightarrow -M^2\, x) \\ P_1^1\, a \,\&\, -P_2^2\, a \\ \hline \therefore\ M^1\, a \,\&\, -M^2\, a \end{array} \qquad \begin{array}{l} (x)(-P_2^1\, x \Rightarrow -M^1\, x) \\ (x)(P_1^2\, x \Rightarrow M^2\, x) \\ -P_2^1\, a \,\&\, P_1^2\, a \\ \hline \therefore\ -M^1\, a \,\&\, M^2\, a \end{array}$$

Case 2:

$$\begin{array}{l} (x)(-P_2^2\, x \Rightarrow -M^2\, x) \\ (x)(P_1^1\, x \Rightarrow M^1\, x) \\ -P_2^2\, a \vee P_1^1\, a \\ \hline \therefore\ M^1\, a \vee -M^2\, a \end{array}$$

Case 3:

$$\begin{array}{l} (x)(P_1^2\, x \Rightarrow M^2\, x) \\ (x)(-P_2^1\, x \Rightarrow -M^1\, x) \\ -P_2^1\, a \vee P_1^2\, a \\ \hline \therefore\ -M^1\, a \vee M^2\, a \end{array}$$

Case 4: This case combines the arguments of Cases 2 and 3, thus allowing the inference of $(M^1\, a \vee -M^2\, a) \,\&\, (-M^1\, a \vee M^2\, a)$, i.e., $(M^1\, a \Leftrightarrow M^2\, a)$. The latter biconditional is sufficient for S_1's and S_2's theories being contradictories. It is also a necessary condition for $M^1\, a$'s and $M^2\, a$'s meaning the same.

tion of such criteria are one factor constituting conceptual change in the development of scientific ideas.

Yet there remain 'analytic truths' which might not be changed in some or perhaps all episodes of the development of a given concept, and these are expressed in the background criteria. By means of these background beliefs or conceptions we may determine what, if any, logical relations are borne by theories that differ in their conceptual content. We can also distinguish between conceptual revision and total conceptual innovation or replacement. In the latter case the criteria and shared background beliefs are insufficient to establish any logical relations of contrarity, contradictoriness, etc. between the theories, leaving them complementary or perhaps even methodologically incommensurable.

4 Syntactical Axioms

I suggested earlier that modifying the syntactical characteristics of an expression is relevant to conceptual change but not sufficient for diagnosing such change. In this section I will expand upon this remark.

An axiomatic statement of a theory, such as classical particle mechanics, non-relativistic quantum mechanics, or classical electromagnetic field theory, should contain among its axioms a number of statements characterizing the nonlogical vocabulary of that theory.[1] These statements, *syntactical axioms*, specify which elements of that vocabulary are predicates, relations, singular terms and functions. They also specify sets of objects over which these expressions are defined. They specify expressions which, when appropriately juxtaposed with certain symbols for the nonlogical expressions, generate further terms or propositions.

For example, an axiomatization of classical particle mechanics might contain statements such as the following:

Every *material particle* has a unique positive real-valued mass.

Symbolically, this statement may be expressed in the form

$$M\colon \Sigma \rightarrow R^{+}$$

which says 'M is a function from the set Σ of material objects onto the set R^{+} of positive real numbers.' Similarly, force is characterized as follows:

Internal force is a vector-valued quantity defined for material particle pairs.

In symbolic terms, $F\colon \Sigma \times \Sigma \rightarrow R^{3}$. That is, F is a function from the cartesian product of sets of material particles Σ^{2} onto a real vector space R^{3}.

[1] For axiomatic formulations of various physical theories see [3]. An axiomatic formulation of classical particle mechanics is also in [19]. See also [14].

Once we have a conceptual background in which 'material particle,' 'time interval of duration Δt,' 'spatial interval of length Δs,' 'interval on the set of real numbers R,' etc., are understood, these axioms characterize sharply the distinct quantities force and mass. 'Material particle' can be characterized in a general way by reference to criteria for 'has mass.' 'Time interval of duration Δt' is characterized in terms of an antecedently understood chronology and chronometry, wherein there appear criteria of varying degrees of generality similar to those described for 'mass'. Topology and geometry are similarly appealed to for spatial intervals. Finally 'real number interval R,' etc., are understood in terms of the formal or logical background of the theory (cf. [4], pp. 52–54, 62, 93–108).

With this background, the above axioms capture several important aspects of the technical and quantitative concepts of mass and force appearing within the classical framework. Mass is construed as a scalar quantity that can take on only positive values. Force, on the other hand, is a vector. That is, direction as well as magnitude is relevant to specifying force-values, and forces can be resolved into components in any direction in space. Mass is an intrinsic property of all material particles in the classical framework, whereas force-values depend upon attributes of material particle pairs.

In the special theory of relativity an object's mass is dependent upon velocity, where velocity must be specified in relation to a specific inertial frame of reference. Accordingly, an object's mass must be specified with respect to an inertial frame. The classical syntactical axiom for mass then is revised to read:

> Every material object in relation to a specific inertial frame has a unique positive real-valued mass.

Thus, in the framework of special relativity mass is a relational and not an intrinsic quality.

5 *Pragmatic Import*

Three factors are relevant to the conceptual content of a theory and must be specified in any complete account of the sense of the non-logical terms of a theory. I have mentioned the criterial statements and the syntactical axioms in which a given expression is characterized. A third factor that I have not mentioned is the overall selection of extra-logical expressions used in a given theory, syntactical axioms and criterial statements. For example, '... has mass' may appear in the extralogical vocabulary of one but not of another theory. I will now attempt to relate these factors in terms of their bearing upon the capacities of individual researchers, viz. their bearing upon strategies and techniques that facilitate, but are not necessary or sufficient for, successful empirical investigation and theory formation. In particular, I am interested in techniques that pose in particular circumstances problems regarding the identification, description and explanation of physical objects and events.

The overall selection of non-logical expressions in a given theory represents a belief on the part of a proponent of that theory regarding which properties, relations and quantities are relevant to the objectives of research within that theory. For example, we might suppose that one objective of research in classical particle mechanics is the explanation of the behavior of macroscopic material systems such as pendulums, planets, and vibrating strings. The selection of vocabulary for classical particle mechanics constitutes a commitment to the view that an empirical evaluation of the number of particulate components of a system, their masses, initial positions and velocities and their interactions, will contribute to the achievement of this objective. On the other hand, the determination of their color, taste, or perhaps their chemical constitution is held to be far less, if at all, likely to contribute to the explanation of their behavior (cf. [17], Section III).

This selection of terms not only directs an investigator to look for specific properties, relations and quantities, but also sets specific limitations within which these properties, etc., can occur. That is, one who subscribes to the syntactical axioms of classical particle mechanics holds the view in advance of any empirical investigation that an object's mass will never be less than zero and can always be represented by a unique positive real number. Similarly, a force must have a specific direction in space as well as a unique real magnitude. Masses of individual particles also are not expected to change in time or with any other mechanical attribute, but forces may vary in time according to specific parameters describing the conditions of interacting objects.

By contrast, a researcher acting in accordance with the syntactical axioms of special relativity believes that he must specify an inertial frame before asking for a unique answer to 'What is the mass of object *a*?' It is thus that the direction of research under the special theory differs from that under classical particle mechanics. Yet the difference is not total, for background criteria shared by the two theories entail that the question is relevant in the same general circumstances, e.g., when confronted with a visible object, one that is not moving with maximal velocity, etc.

Our capacities for carrying out empirical research can be ordered with regard to the relation '... is a prerequisite for ...'. For example, before one can appreciate the fact that the application of '... has mass' constitutes a problem for empirical research, he must believe that the predicate applies in some but not in other circumstances. Also a prerequisite for the problem of evaluating quantitatively the mass of an object is that the qualitative predicate '... has mass' be correctly applied. Accordingly, one must understand and assent to some of the criteria peculiar to '... has mass' (as distinct from other contingent predicates '... is colored,' etc.), viz. one or more of the sufficient conditions for its application, before he can deliberately act in the application of that predicate and, *a fortiori*, before he can legitimately raise a question regarding its quantity. Similarly, an investigator must understand and assent to criteria for the relations '... is greater in mass than ...' and '... is equal in mass to ...' before considering problems such as 'How much greater in mass is σ than σ'?'.

This ordering of prerequisites corresponds to an ordering of criteria regarding theory specificity. This correlation I take as further grounds for my claim that conceptual development is the development of capacities for conducting research.

In the case of mass, visibility, tangibility, etc. appear in common sense as well as all physical theories as sufficient conditions for applying the qualitative predicate '... has mass.' These conditions determine a non-technical sense for the predicate, a sense which is as nearly identical to the sense of '... is a material object' as can be diagnosed in view of the vagaries of 'ordinary language' and commonsense. We may place the necessary and sufficient conditions '... is capable of less than maximal velocity' as adding a more technical aspect to the meaning of '... has mass,' for the addition of this property extends our understanding of mass to objects that do not satisfy conditions of tangibility, visibility, etc. Accordingly, our capacities for applying the predicate are extended to cases in which we believe with some justification, whether empirical or theoretical, that velocity is less than maximal. These capacities are shared by specialists committed to relativistic particle mechanics and by some committed to classical particle mechanics.

However, against this background, relativistic and classical physicists differ in their qualitative and quantitative conceptions of mass. The qualitative conception is significantly clarified by the assumption that maximal velocity coincides with the velocity of light, *c*. The quantitative conception is revised (though not further clarified) by the syntactical assumption relativizing mass to inertial reference frames and by new criteria regarding '+' as an operation for adding mass-values.

6 *The Historical Development of the Concept of Mass*

A final test for my thesis regarding criteria and meaning change is its adequacy in explicating the historical development of the concept of mass. In particular, can my account of meaning change explicate what seems to me to be an incontrovertable feature of conceptual development, viz., its continuity?

In the early stages of development of dynamical concepts, mass and forces of weight and resistance to motion were not sharply distinguished. Aristotle did subscribe to tangibility as a criterion for materiality in his choice of 'haptic' (tangible) qualities, e.g., hot, cold, wet, dry, dense, rare, rough, smooth, hard, soft as fundamental qualities of matter [6]. However, his failure to grasp the inertial aspect of mass is evident in his construal of resistance to motion entirely as a result of interaction with the surrounding medium, and not as an intrinsic quality.

JEAN BURIDAN (c. 1350) characterized material bodies as bearers of impetus, a disposition to continue moving until corrupted by opposing forces. The capacity for impetus depends upon one inherent attribute, the 'quantity of matter,' as well as the actual motion possessed by a material object. However, impetus is distinct from classical inertia in that it manifests itself as a cause and not an effect of motion in empty space, and hence is a force.

KEPLER was probably the first to combine the attributes of inherency and passivity that are shared by the classical conception of mass [14]. His 'inertia' is manifest in the resistance of material bodies such as planets to continued motion, and this resistance depends only upon the inherent 'quantity of matter' possessed by the planet. The only remaining logically relevant feature differentiating KEPLER'S from NEWTON'S quality 'mass' is the Newtonian specification that resistance to continued motion is resistance to acceleration, not velocity or rest. This specification can be construed as a revision of KEPLER'S criteria for '... has mass', an addition to criteria shared by KEPLER and NEWTON whereby mass and force are distinguished, and finally an addition to criteria shared by all regarding which objects are material.

Accordingly, the development of the qualitative conception of mass is incremental, though not always cumulative. The continuity is preserved by shared background criteria, but incremental specifications can be and have been withdrawn.

References

1. ACHINSTEIN, P.: Concepts of science: A philosophical analysis. Baltimore: Johns Hopkins Press 1968.
2. – On the "meaning" of scientific terms. J. Phil. **62**, 266–274 (1965).
3. BUNGE, M.: Scientific research, vols. I and II. Berlin-Heidelberg-New York: Springer 1967b.
4. – Foundations of physics. Berlin-Heidelberg-New York: Springer 1967a.
5. – Physical Axiomatics. Rev. Mod. Phys. **39**, 463–474 (1967c).
6. DIJKSTERHUIS, E.J.: Mechanization of the world picture, p. 22. Oxford University Press 1961.

7. DRETSKE, F.: Seeing and knowing. Chicago: Chicago University Press 1969.
8. FEYERABEND, P. K.: Explanation, reduction and empiricism. Minnesota studies in the philosophy of science, vol. 3, eds. H. Feigl and G. Maxwell. Minneapolis: University of Minnesota Press 1962.
9. FINE, A.: Consistency, derivability and scientific change. J. Phil. **64**, 231–240 (1967).
10. FREGE, G.: On sense and reference. Translations from the philosophical writings of Gottleb Frege, P. Geatch and M. Black. Oxford: Basil Blackwell 1892.
11. GIEDYMIN, J.: The paradox of meaning variance. British J. Philos. Sci. **21**, 257–268 (1970).
12. HANSON, N. R.: Patterns of discovery. Cambridge: Cambridge University Press 1958.
13. HESSE, M. B.: Fine's criteria for meaning change. J. Phil. **65**, 46–52 (1968).
14. JAMMER, M.: Concepts of mass. New York: Harper and Brothers 1961.
15. KLEINER, S. A.: Erotetic logic and the structure of scientific revolution. British J. Philos. Sci. **21**, 149–165 (1970a).
16. – Ontological and terminological commitment and the methodological commensurability of theories. Read at the Philosophy of Science Assoc., second Biennial Meeting (1970b).
17. KUHN, T. S.: The structure of scientific revolutions, second ed. Chicago: Chicago University Press 1970.
18. SHAPERE, D.: The structure of scientific revolutions. Phil. Rev. **73**, 383–394 (1964).
19. – Meaning and scientific change. Mind and cosmos, ed. R. Colodny. Pittsburgh: University of Pittsburgh Press 1966.
20. SUPPES, P.: Introduction to logic, p. 291–304. Princeton: D. van Nostrand 1957.
21. TOULMIN, S. E., GOODFIELD, J.: The architecture of matter. New York: Harper Brothers 1962.

11 The Relevance of Philosophy to Physics

TOM SETTLE

Department of Philosophy, University of Guelph
Guelph, Ont. (Canada)

1 Introduction

My thesis, that philosophy is relevant to physics, is unfashionable, to say the least. This minority view gives me some strange bedfellows: for example, some people with particular religious views who are anxious to make science agree with theology (although primarily their interest is in geology and biology, rather than in physics) [32]. I prefer the company of those philosophers who have been influenced by Sir KARL POPPER'S almost lone defence of the meaningfulness of non-scientific statements – lone, in the sense that POPPER fought in the interests of science [2, 10, 13, 34, 35]. By contrast, the majority view is either a deliberate rejection of philosophy, especially metaphysics, as false (or meaningless or irrelevant), or a simple ignoring of the problems of which philosophy has traditionally been the study. Nevertheless, neither rejecting philosophy nor ignoring it implies that solutions to philosophical problems are not adopted. On the contrary, it is customary for the majority to adopt a stance which is a mixture of various solutions to a number of still puzzling philosophical problems but which interferes as little as possible with scientific work [39]. Furthermore, it is customary to fend off all attempts at serious discussion of those problems in case such discussion should have mischievous effects on the progress of science. This stance is adopted deliberately by some, but implicitly by most. Of course, some metaphysical debates have seriously threatened mischief in science, particularly debates provoked by a clash between theories in science and doctrines in theology; and scientists wish, understandably, to avoid repetition of such affairs [21, 46]. However, the contemporary rejecting (or ignoring) of metaphysics and even of philosophy in general extends the nineteenth century attack on religion in particular in a mischievous manner. Exit the baby with the bath water. In this paper, without disagreeing with the nineteenth century rejection of divine inspiration (if there be any) as a proper source of reliable knowledge (if there be any) in science, I argue in favour of the relevance of the study of philosophical problems for physics.

My thesis that philosophy is relevant to physics is unfashionable not only among physicists but also among philosophers. While physicists are supposedly trying to protect themselves from involvement in what they expect to be barren speculative disputes, by their rejection of relevance, many philosophers appear to be trying, by an analogous rejection, to protect themselves from the impact of what they take to be changing fashions in physics (perhaps under the supposition that only unchangeable truths are proper subjects of inquiry for them) and, what is perhaps worse, some of them even appear to be trying to excuse themselves from knowing about physics at all. These remarks do not apply at all to the originators of a very influential school of thought in contemporary philosophy of science, who were very well versed in current theories of science and who aimed at unifying science. This group of men, with the members of the Vienna Circle as leaders, gave a new impetus and a new style to positivism [6, 7, 16, 17, 22, 24]. Officially, they rejected metaphysics, and identified philosophy with logic and methodology. Actually they engaged in the study of methodological problems only after adopting solutions to problems in metaphysics and other branches of philosophy. For example, it was assumed that there was a real world external to the experimenting scientists; that this real world could be more or less accurately and reliably described in gross (macroscopic) terms; and that unobservable entities might be said to exist only insofar as their effects could be observed. The positivist programme of research was then the logical analysis of physical theory in order to reduce the whole body of scientific knowledge to relations between observable terms. Now, I have no objection to the adoption of a metaphysical stance, as such. I do, however, object to any pretence that positivists were not doing metaphysics. That pretence helped to foster the illusion, now widespread, even in the scientific community, that metaphysics especially and philosophy in general is irrelevant to science. This is a very dangerous illusion since it stifles critical debate concerning the assumptions of science, the canons of scientific practice, and the relations of science as a social institution to society in general. The rejection of philosophy, in short, stifles debate *about* science: science may be practised; apprentice-scientists may be taught their trade (as practised by the masters); the results of science may be used, for better or for worse; but science may not be talked about.

In this paper I hope to give some hint of why science should be talked about by exposing a range of more or less interesting and important philosophical questions.

2 *An Elementary Analysis of Physics*

Physics may be viewed variously as:

P_1: a body of knowledge;
P_2: a discipline of study;
P_3: a social institution.

These ways of viewing physics are not independent since the body of knowledge provides the raw material (background assumptions, problems, etc.) for the individuals following the discipline and grows according to their success, while the individuals in their corporate pursuit of knowledge make up the social institution. Nonetheless, P_1, P_2 and P_3 are distinguishable. Furthermore, they should be distinguished in any discussion of the relevance of philosophy to physics so that it is clearer what is being affirmed or denied, and so that one may assess the importance of a thesis more readily, for it is obviously a weightier matter to assert philosophy is relevant to P_1 (in symbols: $M \xrightarrow{r} P_1$, letting 'M'—for metaphysics—stand for philosophy, since 'P' is already in use) than it is to assert philosophy is relevant to P_3, since $M \xrightarrow{r} P_1$ implies $M \xrightarrow{r} P_3$, but not *vice versa*. (In some cases, $M \xrightarrow{r} P_2$ implies $M \xrightarrow{r} P_1$, for example if philosophical considerations were to clarify problems of method which in turn led to changes in P_1; but in other cases, not, for example if philosophical considerations were relevant merely to the state of mind of the research worker, giving him a greater understanding of what he was doing or enhancing his incentive to subject himself to the discipline. Even so, he might work more successfully as a result and thus affect P_1.)

I aim to show not only that $M \xrightarrow{r} P_1$, but also, and even independently, $M \xrightarrow{r} P_2$, and $M \xrightarrow{r} P_3$. It is by no means my concern to show that *all* philosophy is relevant to physics—on the contrary, I think a great deal of philosophical discussion is quite irrelevant—but only to show that some philosophy is relevant to physics in all its aspects.

Since 'relevance' names a key concept in my thesis, it behoves me to give at least a cursory account of the meaning of the term. The notion of relevance is tricky since it involves the notion of propensity—which has not yet been satisfactorily elucidated in the literature—and the notion of making a difference, which is a sort of causal concept. Operationally, we could say X is relevant to Y if and only if a change in X (ΔX) over an interval of time Δt implies a change in $Y(\Delta Y)$ over some corresponding interval $\Delta t'$:

$$X \xrightarrow{r} Y =_{\mathrm{df}} (\Delta X \neq 0)_{\Delta t} \rightarrow (\Delta Y \neq 0)_{\Delta t'}.$$

Superficially this definition is satisfactory since we should certainly want to say that if changes in X lead to changes in Y, X is relevant to Y. But X may be relevant to Y even if some changes in X do not result in changes in Y and even if neither X nor Y changes. In this latter case we might want to say that a change in X might cause a change in Y. Thus I should prefer to say, X is relevant to Y if and only if changes in X have a propensity to lead to changes in Y. (*Prop* for propensity):

$$X \xrightarrow{r} Y =_{\mathrm{df}} Prop[(\Delta X \neq 0)_{\Delta t} \rightarrow (\Delta Y \neq 0)_{\Delta t'}] \neq 0.$$

Perhaps this is as deep an analysis as is needed for the purposes of this paper.

3 *An Elementary Analysis of Philosophy*

I propose to view philosophy, primilarly, as a discipline of study, and to analyse it according to the problems under study. Consequently, I shall also have in view the aspect of philosophy as a body of knowledge, namely, as a variety of answers to philosophical problems. It is not my intention to argue either way on the question of the relevance to physics of philosophy as a social institution. I wish to engage my readers' attention, specifically, in inquiry into the problem of the relevance to physics of the study of some philosophical problems.

Philosophical problems cluster in five main groups round central questions:

1. What is there, basically?
2. How do we know what there is?
3. How do we learn what there is to know?
4. How is what we know rationally ordered?
5. What values (and dysvalues) are conserved or created by our pursuit of knowledge?

corresponding to five divisions of philosophy:

M_1: Ontology (Greek *ontos:* being).
M_2: Epistemology (Greek *episteme:* knowledge).
M_3: Methodology (Greek *methodos:* mode of inquiry).
M_4: Logic.
M_5: Axiology (Greek *axios:* valuable) (including ethics and aesthetics).

Let us consider these divisions briefly, in turn.

4 *Ontology*

Consider physics as a body of knowledge (P_1). Baldly, physics is then a set of sentences, some specific, some general, some in ordinary languages, some in technical symbols, containing terms which name entities or properties of entities, and express relations between them. In a grossly oversimplified fashion, we might represent physics as a set of sentences of type (i):

$$\text{(i)} \quad Fa$$

where 'a' names an object and 'F' a property; and a set of sentences of type (ii):

$$\text{(ii)} \quad (x)_A(Fx)$$

where 'x' is a variable naming members of a class A. The entities and properties named may be deemed observable or unobservable. Empirical laws in physics usually generalize observable properties of observable entities, while theories proper may include unobservable (perhaps fictitious) properties of observable entities and even unobservable (perhaps fictitious) entities.

Consider physics as a discipline (P_2). It is usually understood (perhaps incorrectly) that an aim of physics is to say what sort of things the universe is made of and what the properties of these things are. This is one form of the central ontological question. As disciplines, ontology and physics share direction and aim. Hence we should be able to view P_1 as a subclass of ontology, viewed as a body of knowledge. This should not be a surprising result, since, historically, some ontological theories have been early forms of physical theories. Some ontology not yet physics may be physics of the future. From this point of view, ontology is certainly relevant to physics. In addition we might view ontological theories as frameworks for the development of physical theories and for their interpretation. All this has been very nicely discussed by AGASSI [2].

An example of an unresolved ontological dispute which is not without significance in physics is the problem of the deterministic character of the universe. For example, it is generally agreed that probabilities are involved in quantum theory but it remains unsolved what is being said about the world when this is agreed. Is the world at least partially indeterministic, in fact? That is, is chance objective? Or is it simply the case that we have not yet found the deterministic laws underlying quantum theory? Or should we understand the probabilities as always associated with our ignorance? [12, 23, 36, 45]. In addition to its interpretative significance any metaphysical (ontological) framework used to interpret physical theory will be significant in directing the course of future research since it will pose problems worth trying to solve which will be different from the problems posed by rival ontologies. Hence $M_1 \xrightarrow{r} P_1$ and $M_1 \xrightarrow{r} P_2$.

A second domain of relevance is the debate over what physical theories are about [14, 30]. Two questions are usually confused here: the question whether terms in physical theories are intended to refer genuinely to the physical world or are merely ficittious, and the question whether the real world or only the world of appearances is accessible to scientific inquiry. The history of the confusion is interesting but I shall not retell it here. More recently the confusion has been compounded by the distinction between observables and unobservables. Let T be the class of terms occurring in theories in physics; let G be the sub-class of terms thought to refer genuinely; F the sub-class thought to be fictitious; let R be the sub-class referring to the real world; A to the apparent world; let O be the sub-class of terms naming observables; U, unobservables; such that:

$$G \cap F = \emptyset; \quad R \cap A = \emptyset; \quad 0 \cap U = \emptyset,$$

then

$$T = G \cup F; \quad T = R \cup A; \quad T = 0 \cup U.$$

Theories about the manner in which the sub-class boundaries are related to each other are many and varied. One of the earliest is naive realism, which may perhaps be presented as:

$$U = F = A = \emptyset, \quad \text{hence } T = O = G = R. \tag{1}$$

Naive realism, the theory that the world really is just what it appears to be on careful inspection by a sober, alert, intelligent (etc.) observer, is a very plausible theory. It was not significantly undermined by such arguments as those presented by DESCARTES [19] as to why we should doubt the veracity of perceptions. Nonetheless a major problem for philosophers of the seventeenth and eighteenth centuries was to account for empirical knowledge. Furthermore this problem was never solved. I discuss it further in the next section. Here it suffices to note that the problem what physical theories might properly be said to be about was made more difficult by a justificationary constraint. Thus, if $(x)_A(Px)$ represents a law in physics, then, it was thought A could refer genuinely to the real world only if the members of A were knowable, in the sense that it may be known that they exist, and hence, what some of their properties are. (If the members of A are not knowable, then the reference is fictitious.) But members of A were thought to be knowable only if their existence (and hence some of their properties) could be proved or, in some strong sense, justified. KANT thought that this latter requirement could not be met for the real world (that is, for any thing-in-itself), hence he dropped the idea that $G \cap R \neq \emptyset$ [26].

A further distinction was that between concepts which were supposed to be derived from experience and those which were invented (conjectured). Let C be the class of conjectured terms in physical theories and I the class of terms inferred from experience (by induction) so that $T = C \cup I$ and $C \cap I = \emptyset$. Further let J be the class of terms we are (whether inductively or not) justified in using, while S is the class of purely speculative terms or of terms used in a purely speculative fashion, such that $T = J \cup S$ and $J \cap S = \emptyset$. Then we may distinguish DESCARTES's intellectualist view, (2), from BACON's inductivist view, (3) [18, 19, 8]:

$$S=\emptyset; \quad I=\emptyset; \quad C=J=G=R, \tag{2}$$

$$S=\emptyset; \quad C=\emptyset; \quad I=J=G=R. \tag{3}$$

KANT's theory to which I have already referred is perhaps:

$$R=\emptyset; \quad C\cup I=J=G=A. \tag{4}$$

Positivists reacted against KANT's pessimism about knowledge of the real world without dropping the justificationary constraint. They identified what could not be known with what could not be observed [9, 17]. Their view may be represented thus:

$$O=J=G=R; \quad U=S=F\neq\emptyset. \tag{5}$$

With their view that justified physical theories might contain fictitious and unjustified terms they introduced an interesting note of corrigibility into the theory of physical theories; and, in addition, *re*introduced instrumentalism, the theory that physical theories are *merely* instruments for prediction, hence it is a matter of indifference what terms are used as long as the phenomena are saved.

POPPER rejected instrumentalism, but went further than the positivists in loosening the justificationary constraint. POPPER suggested that it be dropped altogether: all knowledge was to be thought corrigible [34, 35]. Under POPPER's view, unobservable terms may be genuine and observable terms may be unjustified. Hence POPPER allowed for a more sophisticated version of realism. His view appears to be:

$$A=\emptyset; \quad I=\emptyset; \quad J=\emptyset; \quad \text{hence } T=R=C=S; \quad G\cap U\neq\emptyset. \tag{6}$$

This is the theory I endorse (conjecturo-critical realism). The question of occult qualities will be taken up again later.

The major realist component of (6): $A=\emptyset$, may be thought to imply that no phenomenal terms enter physical theory. Terms referring to actual objects or to their apparent properties (colour, texture, etc.) do occur in physics, of course, but they occur in reports of observations or in instructions in experimental physics, rather than in physical theory proper. Hence, P_1 contains sentences in which phenomenal terms occur, while T does not. Whenever a theory is to be tested, part of the work to be done in preparing it for the test is to supplement the theory with assumptions enabling statements about observations to be deduced. In reply to KANT, we may perhaps urge that theoretical physics, which represents symbolically a possible physical world, is intended to represent more or less closely the real world and thus to serve as an explanation of appearances which differ from the postulated real world considerably. Furthermore, although we cannot prove that the entities we posit exist (or have the properties we ascribe to them), our supposition that they do enables us to explain appearances in a non-arbitrary fashion, and is corrigible under criticism, including empirical criticism.

I contend that different views on the question what physical theories are about introduce into P_1 different interpretations of the symbols used in theoretical physics and hence, again, $M_1 \xrightarrow{r} P_1$. Furthermore, these different views affect the account we give of the discipline. Hence $M_1 \xrightarrow{r} P_2$.

5 *Epistemology*

In the Western intellectual tradition it has customarily been supposed that a sufficiently large number of fundamental truths are either self-evident, or may readily be proved. Any rational man (obviously the best kind to be) would hold all and only those views which were self-evident or demonstrable. By contrast, on the one hand, the dogmatist was thought to be the man who held tenaciously to what could not be proved (or even to what could be disproved) while, on the other hand, the skeptic was thought to be the man who stubbornly refused to accept what "rational" men thought obvious or demonstrable. In what follows, I depart from the Western tradition and advert to skepticism, of a rather special kind.

The epistemological problem raised by the challenge offered by early men of science (COPERNICUS, GALILEO, DESCARTES, BOYLE, NEWTON, etc.) to establish learning was 'How do we know what there is to know?' As might have been expected, given the rational tradition, this question was, more specifically: 'What warrant can we give, what justification, for what we claim to know?' Alternatively 'What authority may we cite…?' Two authorities vied for the role: the intellect and experience. It did not take long in the modern period for it to be clear that the intellect would not do as authority in science. On the other hand, scientists and philosophers still have not quite given up hope for experience as the authority to follow. Moreover, the theory that experience is a proven authority in matters of fact has become, after 300 years of success in science and technology, a matter of common sense. Consequently it is much harder to criticize and to replace. Nonetheless a close examination shows that experience will not do, either. Here are three devastating arguments against the theory that experience should be regarded as a final and authoritative arbiter on matters of fact:

(1) Experience cannot tell us that what we take to be the laws of nature will not alter in the future (say, tonight) (MAIMONIDES) [29].

(2) Experience is insufficient warrant for a statement reporting experience, since we may be deluded, dreaming, drunk, or simply mistaken in our judgment of what we experienced (DESCARTES) [19].

(3) No universal statement is ever proved by the accumulation of particular statements which are instances of it (unless the instances exhaust the possibilities) (ARISTOTLE) [5].

Let us put these objections another way. Three steps are required for experience to prove scientific theories to be true:

(i) Experience (E) has to warrant observation reports (O):

(ii) Observation reports have to warrant laws as we think them to be (L).

(iii) The laws (L) have to be identified with true laws of nature (T).

Since members of L are intended to be invariant over time and space but members of T may not be, the third step is not justified even if the first two are. It is an ontological problem whether the universe persists with the same behaviour patterns. Traditionally, believers in a powerful deity have argued for the view that the deity is free not only to intervene in a manner which does not violate T (naturalist theory of miracles), nor only, in addition, to intervene in a manner which does violate T, but also he is free to alter T, say to T'. But belief in a powerful deity is not a necessary condition for belief in the possibility that T may alter.

Usually we set MAIMONIDES' objection aside. In doing so, we may say that though our aim may be to improve L so that it approximates T more closely, the only way we can do this is to aim for a version of L justified by O which in turn has been justified by E; if T alters (to T') then in time E will alter to E' and O to O'; in our efforts to explain O', we should change L to L' hopefully approximating T'. But we should be wrong in replying to MAIMONIDES this way. If we changed L to L' because of changes in E and O subsequent upon altering T, L' would not be approximating T' so much as a superordinate set of laws T^* of which T and T' were special cases, since in inventing L' we would be trying to accommodate *both* E and E' in laws that were invariant with respect to time and space. Hence we should never notice if the true laws of nature changed, since the resultant changes in E would be indistinguishable from those we might have expected if L was still only a poor approximation to T. We would mistakenly think of L' as a better approximation to T. Nonetheless, for the sake of progress in science, we do set MAIMONIDES' objection aside. This should perhaps be regarded as a *methodological* decision: it sets one of the rules for science.

DESCARTES's objection has received much more attention from philosophers in modern times than has MAIMONIDES'. Even so, it is usually ignored by scientists who agree that if enough of us do research and check each other's results we can overcome the risk of being mistaken in reporting what we observe, at least on the macroscopic scale. This is a pragmatically sufficient rejoinder to DESCARTES's objections and it has led to recent lack of interest among philosophers in the problems of the empirical basis of science. Nonetheless, it is not a sufficient rejoinder to another version of DESCARTES's objection, a version raised by POPPER, namely that since all observational terms are general terms they are theory-laden [34]. Now, if theories are to be proved by induction from experience, all inductive proofs (if there be any) will suffer from a mild attack of *petitio principii*. (Recently, a similar point has been made by GOODMAN.) The importance of the problem lies here: the names used in theories in science indicate classes, thus dividing objects in the real world into these classes. Perhaps the divisions thus rendered are natural, perhaps not. Complacency over terminology may lead to stagnation, while a lively sense of the possible inadequacy of the classification produced by physical theory may encourage criticism and improved conjecture. Thus in this respect $M_2 \xrightarrow{r} P_2$ and possibly also $M_2 \xrightarrow{r} P_1$.

ARISTOTLE's objection, which is repeated by HUME (whose presentation confuses it with MAIMONIDES') [25], cannot be at all easily set aside. There clearly are logical relations of deducibility between observational (particular) data and general laws: laws entail data; some data refute laws. In symbols:

$$\text{if } (x)_A(Fx) \quad \text{and} \quad a \in A \text{ then } Fa;$$

and

$$\text{if } b \in A \quad \text{and} \quad \sim Fb \text{ then } \sim (x)_A(Fx).$$

On the other hand, the inductive relation is less firm:

$$(x)_A(Fx)$$

is *not* entailed by:

$$Fa \,\&\, Fb \,\&\, Fc \,\&\, Fd \ldots$$

$$(\text{where } a, b, c, d \ldots \in A)$$

unless the class A is exhausted by the instances cited.

The situation is perhaps even worse than this, since, as DUHEM pointed out long ago [21], there is an infinite number of mathematical functions which "fit" any finite number of observation reports. He thought this introduced considerable arbitrariness into the construction of physical theory. (Later I suggest that arbitrariness is constrained by aesthetic desiderata.) With the admission of the failure of full justification of scientific laws, it was natural, I suppose, that philosophers and scientists should look for as strong a justification as possible. Despite HUME's arguments against trying even to probability laws [25], some interpretation of the Probability Calculus was sought [16] (and in some now dwindling circles, still is) under which there could be measured the degree of confirmation an hypothesis gained on certain evidence. Many philosophers saw the power of negative instances in refuting generalities, contrasted to the ineffectiveness of positive instances (even of varied kinds) in lending support, but the first to propose criticism (refutation) as the key to the non-arbitrariness of hypotheses in science was POPPER. POPPER correctly saw that confirmation or justification, in the sense of the accumulation and parade of positive instances, did not distinguish science from pseudoscience. He proposed that empirical falsifiability be the criterion for the scientific status of strictly universal sentences (laws); and he further proposed that science could be distinguished from non-science (especially pseudo-science) by a *critical policy*, the policy of submitting our hypotheses to servere testing in the hope of uncovering their faults. (The problem remains, however, which of the various theories, models, laws and specific statements assumed in making a prediction should be taken as false when the prediction is refuted under test: this is a very interesting methodological problem to which, I think, there is no solely syntactical solution. I refer to it again below).

The proposal of POPPER's, to give up trying to justify laws and to try to refute them instead, is so revolutionary it seems sometimes not to be understood even by some of his admirers. For instance, LAKATOS appears to think failure to refute equals partial justification [27]. POPPER's philosophical opponents certainly do not comprehend his proposal ([37, 38] cf. [43]; [20] cf. [40])—which is not surprising, since it would impose a radically different programme of research upon philosophers and even upon scientists. Traditionally, in epistemology, philosophers have sought a formula for showing that this or that method of arguing or of introducing evidence, or this or that authority or source of information, was the proper mode or source to use in establishing what was being claimed, so that all reasonable men would be *obliged* to concur. This happy formula has never been found. In its place, POPPER has proposed critical debate. An old enough suggestion, to be sure. "Back to the Presocratics", indeed! Everyone knows one must use critical debate. POPPER's proposal however goes much further than is customary. His proposal is not simply to use critical debate as a prolegomenon to proof, as a way of dealing with the opposition before establishing one's own position, but rather as the whole mode of inquiry. Traditionally, knowledge has been viewed as justified true belief; POPPER proposes to view it as not-yet-refuted, possibly false, conjecture.

The relevance for physics of the debate over epistemology need hardly be argued: P_1 includes observational reports, empirical laws, and theories which have passed some criterion of adequacy, else just any report or putative law would be included; epistemology is the study of some of these criteria of adequacy. Shall we include in P_1 only those statements for which we can give warrant, and if so, how strict is our notion of warrant? Or shall we, on the other hand, include theories which, although we cannot give them warrant, nonetheless satisfy the criterion of explaining certain puzzles, or solving certain problems, to which experience has given rise? When (if ever) shall we remove an erstwhile satisfactory theory from P_1 and on what grounds? These questions are specific variants of the central epistemological question and they clearly are relevant to physics, at least because varying answers may vary the content of P_1. Hence, again, $M_2 \xrightarrow{r} P_1$ as well as $M_2 \xrightarrow{r} P_2$.

What should we now say about the credence to be placed in claims about the existence of occult entities or qualities? Rejection of the occult, either hidden properties of visible things, or hidden things, follows from the requirement to justify knowledge empirically. In modern times it seems to have been thought that a necessary condition for the adoption of occult entities or qualities is the rejection of empiricism. This is not so. The rejection of the requirement to justify knowledge is sufficient to liberate us to adopt hypotheses or theories positing the occult. The rejection of the requirement to justify scientific knowledge empirically might, however, be thought to threaten the empirical basis of science, since traditionally this has been thought to be secured by the mass of empirical evidence in favour of the laws and theories of science. POPPER has suggested the alternative that the empirical basis of science is secured by the empirical testability of scientific laws. Hence occult entities or qualities may be admitted, if hypotheses about them could be rendered testable, even if only indirectly. In my view, we may go even further, taking our cue from POPPER's definition of the aim of science: "to find satisfactory explanations of whatever strikes us as being in need of an explanation", ([33], p. 24) and say that occult entities and qualities may be properly introduced into scientific discourse if they help explain what we think needs explaining. The general requirement that we hold our explanations open to criticism (including empirical criticisms, if any can be devised) is, it seems to me, a sufficient constraint upon arbitrariness.

Perhaps it is only a question of emphasis, perhaps it is more fundamental, that POPPER's view shifts our attention from support to explanation. Logically these are complementary: generalities explain instances, instances support generalities. I suspect the shift is of some importance, however, not only logically since explanation is deductive while support is inductive, but also methodologically since a stress upon explanation will encourage growth of knowledge by encouraging improvement in explanations while a stress on support will discourage growth by encouraging entrenchment. Hence, again $M_2 \xrightarrow{r} P_1$ and $M_2 \xrightarrow{r} P_2$.

6 *Methodology, Logic, Axiology*

The methodological question, How do we learn what there is to know?, cannot be separated from questions in the logic, ethics and aesthetics of science, since methodology studies the means by which certain ends are sought: the means are supposed to be rational, hence logical considerations enter; and any consideration of ends and means implies value theory. This is especially the case in view of the discussion in previous sections. Let me elaborate.

In the days when it was assumed that truth was easy to come by, that thorough justification could be given for knowledge, and that anybody who applied himself diligently might learn all there was to know (or at least, all there was that was worth knowing) it was plausible to suppose that knowing was beneficial. Hence, scientific inquiry justified itself by its success. Nowadays all these assumptions are questioned, as is even the criterion of success. There is too much to know for one man to know all, hence he must choose in which area to seek education and to prosecute research. This raises any number of moral and aesthetic considerations for the individual. Furthermore, knowledge is not thought now to be incorrigible. It may even be somewhat elusive. Justification of a thorough kind cannot be given for theoretical knowledge, though we may say why we think we are right. (Incidentally, the less precise and the less general our knowledge is, the more firm it may be regarded, which is why refuted theories such as NEWTON's gravitational theory or BOYLE's Law are firmly entrenched once the limits within which they are held to apply are established. Any theory at all may be established by weakening the demands of accuracy, and any theory refuted by strengthening them.) Saying why we think we are right amounts to explaining how the knowledge we claim to have satisfies certain desiderata. Perhaps in addition we might try to say why these particular desiderata (or criteria of admission to P_1), and not some others, ought to be satisfied. In the nature of the case, then, decisions to include or exclude an hypothesis will have an aesthetic and even a moral component. That some lines of research are more obviously beneficial to mankind, if successfully pursued, than others enhances the importance of the moral component [44].

Let us assume that the quest for truth (or for knowledge, as an approximation to truth) is worth prosecuting, though we may have reservations about whether searching for just *any* knowledge is worth the cost. Then we may ask what criteria we shall set up for judging whether a certain hypothesis is preferable as an approximation to the truth than a rival; and we may ask by what means we might seek even more satisfactory hypotheses. We may also ask what the criteria are which an hypothesis must satisfy for us even to entertain it as a possibility. The answers to these questions will pick out a methodology. Here are a few answers:

(1) In physics, we do not entertain hypotheses which explain nothing.

(2) We count on explanation E_1 of an event e_1 as satisfactory in an elementary fashion if e_1 is implied by E_1, though perhaps not directly. For example: $(x)_A(Fx)$ explains Fa, if $a \in A$.

(3) We count E_1 better than E_2 if E_1 explains all that E_2 explains and, in addition, explains some events E_2 does not explain, especially if E_1 explains events which clash with E_2.

Examples of other desiderata are elegance, simplicity, consistency, and exactness. A requirement which is frequently effective (usually implicitly), and which may be seriously questioned as to its propriety, is agreement with a world-view. Perhaps our hope of improving our world-view lies in seriously entertaining hypotheses which challenge it.

Just as there can be no thoroughly justified scientific knowledge so, analogously, there cannot be any self-evident aesthetic and moral criteria by which to settle disputes as to which hypotheses to entertain, which to adopt, and so on. What characteristics a theory of physics should possess is not immediately obvious. Various champions of various philosophical (religious or atheistic) causes have proposed varieties of sets of characteristics. At root the debate may be ontological: it may be a debate about the characteristics of the universe in general, and hence the characteristics any putative description of the world should have. For example, if there is no objective chance, laws must be deterministic rather than statistical. On the other hand, if the aim of inquiry is thought to be the facilitation of prediction and control of the future, ease of computation and closeness of fit to data will be prime desiderata. And so on. On the surface, however, the debate concerns what features we prefer theories to possess, and consequently, the considerations are aesthetic. POPPER has stressed the conventional character of methodological decisions [34]; and AGASSI has stressed the institutional context of the debate, and subsequent decisions ([3, 4]; cf. [41, 45]). I wish here to stress the aesthetic and moral quality of the desiderata and hence of the decisions. Even should there be a consensus within the scientific community over which desiderata to use and which hypotheses to choose as a result, this would not remove the aesthetic and moral dimension. Incidentally it is unlikely that we could get a general consensus as to the boundaries of the scientific community – all pseudo-scientists as well as all scientists would no doubt wish to draw the boundary so as to include themselves, and until the boundary is drawn we do not know whom to count as competent, as a scientist, to draw it. Vicious circle! Perhaps we should recognize that quite private sets of preferences may motivate the research interest of and the criteria of acceptability used by, individual scientists. Debate between scientists protects the community from arbitrariness but not from (sometimes acrimonious) dispute.

Considerable work has been done in the literature to analyze some of the desiderata. And this is to be welcomed, provided it is not part of a general programme to eliminate aesthetic and moral considerations. Such a programme would be misguided, since, although these considerations may be taken for granted or ignored, they cannot be eliminated. Worse, it would be mischievous, since, without openness there would likely be no criticism; and without criticism errors have too good a chance to be perpetuated – methodological errors as much as errors in theory. On the other hand, a scrutiny of the desiderata that are or may be or (perhaps) should be used, the scrutiny not necessarily including an analysis, in case the desideratum is simple or primitive, but definitely including a comparison of the likely outcomes of the use or non-use of the various desiderata, would improve the chances of the desiderata being appropriate and enhance the competence with which they were used. Hence $M_5 \xrightarrow{r} P_1$ and $M_5 \xrightarrow{r} P_2$. Such a scrutiny would also supply us with a much needed theory of success to replace the traditional, but refuted, theory that success is being right or very nearly right. So much for the aesthetic considerations.

The moral considerations arise mostly from a clash between the interest of individual scientists (and even of groups of scientists) on the one hand and the interests of society in general on the other. This clash is a recent phenomenon: only recently has large scale public funding been made available to science for teaching and research; and only recently have the outcomes of research (in the shape of technological innovations) been recognized so widely as potentially destructive of the quality of life and even of life itself. It is not surprising, then, that the modern image of the scientist as a lonely research worker pursuing truth for its own sake, financed by himself or by some lay benefactor [28] has to be changed. Society has too many pressing problems requiring solutions for it to tolerate the expenditure of all its research funds on projects which match the research interests and curiosity of individual scientists without having any foreseeable beneficial consequences for society as a whole. There is a strong moral case for scientists choosing research problems which relate to social problems and social goals. In this way, moral considerations might influence physics in all its aspects. Clearly if aesthetic and moral considerations are relevant to physics, the study of value as such (value theory in its generic sense) is relevant also. Hence, $M_5 \xrightarrow{r} P$.

Now for some remarks about logic.

The virtues of axiomatization of physical theory have been "preached" persuasively, as well as demonstrated in practice, by Bunge [11]. I concur with his conclusion that exposing the deductively related core of a theory is beneficial to physics, bringing conceptual clarity, stimulating criticism and hence promoting growth of knowledge. Hence $M_4 \xrightarrow{r} P_1$. Similarly, benefit accrues from exposing the deductive logical relations between theories, empirical laws and specific consequences of these. The methodology of science is clearer when the arguments used in criticism are sharpened in focus. For example, it was assumed by many philosophers that negative instances counted in a straightforward fashion against generalities. Duhem seems to have been the first to point out that a battery of assumptions is needed to augment a physical theory before testable consequences may be deduced from it, hence the falsification of a prediction entails the falsity of the conjunction of many assumptions rather than the falsity of any one [21].

A decision is required as to which of the assumptions to take as falsified. A very clear presentation of normal scientific practice, from which, no doubt, there are well argued exceptions, has been given recently by Bunge [13, 16]. I am in favour of an attempt to state what current practice is. This is too frequently ignored by philosophers who address themselves to science as a logical possibility rather than a social phenomenon. Nonetheless, it is appropriate for us to note that, since the considerations which lead to decisions of the kind we are here concerned with are ontological and aesthetic (and perhaps even moral) it might be beneficial to science for this not to be concealed. And beneficial, too, for current practice to be criticized.

By contrast with the fruits of axiomatization, the anticipated fruits were not born of the labours to provide a logic of *induction*. Much was gained by way of clarification of side issues, but the central task of providing an algorism for computing the degree to which an hypothesis was supported by evidence or at least of providing an analysis of the notion of degree of support (or confirmation) appears to be futile [42]. Nevertheless the language of support, the language of a justificationist theory of the relation of evidence to hypothesis is, regrettably, common parlance among scientists. Logical analysis has not yet had its full impact upon the modes of speech, and hence, one might guess, upon the research styles and aims, of many scientists.

In my view, the only tenable theories of method are those which prescribe critical debate not only of theories but also of the desiderata they are supposed to satisfy and which stress the logical connections between the various constituents of P_1. From the unsatisfactory methodology of Bacon we may borrow the notion of experience supplying us both with problems and with criticisms of theories. From the unsatisfactory methodology of Descartes we may borrow the notion of the intellect supplying general concepts and hypotheses to explain experience. Thus what were originally seen as authorities to bind our allegiance (experience and the intellect) are now to be viewed as sources of (corrigible) knowledge, held open to criticism. Science is sometimes represented as a sophisticated version of learning from experience, but this unbalances the mixture. In science we learn not only from experience but also from intuition: the various desiderata which we oblige physical theory to satisfy are not found in the phenomena, but are imposed upon them by us. Provided such imposition is done critically, it is not improper. These methodological considerations are clearly relevant to physics as a discipline: $M_3 \xrightarrow{r} P_2$. I conjecture that a critical methodology used by a competent scientist, with ordinary luck in his research, will improve the body of knowledge: $M_3 \xrightarrow{r} P_1$.

7 *Conclusion*

An analysis of philosophy and a scrutiny of its parts leads us to the view that the study of a quite broad variety of philosophical problems not only is relevant to physics but also may be beneficial to physics as a body of knowledge, as a discipline and as a social institution.

References

1. Agassi, J.: The confusion between physics and metaphysics in the standard histories of science. Ithaca **26**, 231–244 (1962).
2. – The nature of scientific problems and their roots in metaphysics (M. Bunge, ed.). The critical approach to science and philosophy. London 1964.
3. – The confusion between science and technology in standard philosophies of science. Technology and Culture 7, 348–366 (1966).
4. – Positive evidence in science and technology. Philos. Sci. **37**, 261–270 (1970).
5. Aristotle: Prior analytics.
6. Ayer, Sir A. J.: Language, truth and logic. London 1936.
7. – (ed.): Logical positivism. New York 1959.
8. Bacon, Sir Francis: Novum Organon.
9. Bridgmann, P. W.: The nature of physical theory. Princeton 1936.
10. Bunge, M.: Metascientific queries. Springfield, Ill. 1959.
11. – Physical axiomatics. Rev. Modern Phys. **39**, 463–474 (1967).
12. – (ed.): Quantum theory and reality. New York 1968.
13. – Scientific research. New York 1968.
14. – What are physical theories about? Amer. Philos. Quart. Monograph No. 3, 61–99 (1969).
15. – Theory meets experience. Mind, science and history. Albany, New York 1970.
16. Carnap, R.: Logical foundations of probability. Chicago 1950.
17. – Philosophical foundations of physics. New York 1966.
18. Descartes, R.: Discours de la Méthode.
19. – Méditations.

20. DEUTSCHER, M.: Popper's problem of an empirical basis. Austral. J. Philos. **46**, 277–288 (1968).
21. DUHEM, P.: La Théorie Physique; Son Objet, Sa Structure. Paris 1907.
22. FEIGL, H., BRODBECK, M., (eds): Readings in the philosophy of science. New York 1953.
23. GOOD, I.J.: Kinds of probability. Science **129**, 443–447 (1959).
24. HEMPEL, C.: Aspects of scientific explanation. New York 1965.
25. HUME, D.: Inquiry concerning human understanding.
26. KANT, I.: Kritik der reinen Vernunft
27. LAKATOS, I.: Changes in the problem of inductive logic. In: I. Lakatos (ed.), The problem of inductive logic. Amsterdam 1968.
28. LEWIS, S.: Arrowsmith.
29. MAIMONIDES, M.: The guide for the perplexed.
30. MAXWELL, G.: The ontological status of theoretical entities. In: H. Feigl and G. Maxwell (eds.), Minnesota studies in the philosophy of science III. Minneapolis 1962.
31. NEURATH, O., CARNAP, R., MORRIS, C., (eds.): International encyclopedia of unified science. Chicago 1939.
32. PETER, W.G., III: Fundamentalist scientists oppose Darwinian evolution. BioScience, **20**, 1067–1069 (1970).
33. POPPER, Sir K.R.: The aim of science. Ratio **1** (1957).
34. – The logic of scientific discovery. London 1959.
35. – Conjecture and refutations. London 1963.
36. – The propensity interpretation of probability. British J. Philos. Sci. **10**, 25–42 (1959).
37. SALMON, W.C.: The foundations of scientific inference. Pittsburgh 1966.
38. – Justification of inductive rules of inference. In: I. Lakatos (ed.), The problem of inductive logic. Amsterdam 1968.
39. SETTLE, T.W.: Scientists: Priests of pseudo-certainty of prophets of inquiry? Science Forum, **2**, No. 3, 22–24 (1969).
40. – Deutscher's problem is not Popper's problem. Austral. J. Philos. **47**, 216–219 (1969).
41. – The point of positive evidence: Reply to Professor Feyerabend. British J. Philos. Sci. **20**, 352–355 (1969).
42. – Confirmation as a probability: Dead but it won't lie down. British J. Philos. Sci. **21**, 200–201 (1970).
43. – Is corroboration a non-demonstrative form of inference? Ratio (forthcoming).
44. – Review of R. Dubos, Reason awake: Science for man and N. Calder, Technopolis: the social control of the uses of science. In: Science Forum **18**, 18–21 (1970).
45. – Induction and probability unfused. In: P.A. Schilpp (ed.), The philosophy of Karl R. Popper (forthcoming).
46. WHITE, A.D.: A history of the warfare of science with theology. New York 1960.